T0215134

SpringerBriefs in Applied Sciences and Technology

Computational Intelligence

Series editor
Janusz Kacprzyk, Warsaw, Poland

For further volumes:
http://www.springer.com/series/10618

About this Series

The series "Studies in Computational Intelligence" (SCI) publishes new developments and advances in the various areas of computational intelligence—quickly and with a high quality. The intent is to cover the theory, applications, and design methods of computational intelligence, as embedded in the fields of engineering, computer science, physics and life sciences, as well as the methodologies behind them. The series contains monographs, lecture notes and edited volumes in computational intelligence spanning the areas of neural networks, connectionist systems, genetic algorithms, evolutionary computation, artificial intelligence, cellular automata, self-organizing systems, soft computing, fuzzy systems, and hybrid intelligent systems. Of particular value to both the contributors and the readership are the short publication time frame and the worldwide distribution, which enable both wide and rapid dissemination of research output.

Rajesh M. Bodade · Sanjay N. Talbar

Iris Analysis for Biometric Recognition Systems

 Springer

Rajesh M. Bodade
Faculty of Communication Engineering
Military College of Telecommunication
   Engineering
Mhow
Madhya Pradesh
India

Sanjay N. Talbar
Electronics and Telecommunications
   Engineering
Shri Guru Gobind Singhji Institute of
   Engineering and Technology
Nanded
Maharashtra
India

ISSN 2191-530X   ISSN 2191-5318   (electronic)
ISBN 978-81-322-1852-4   ISBN 978-81-322-1853-1   (eBook)
DOI 10.1007/978-81-322-1853-1
Springer New Delhi Heidelberg New York Dordrecht London

Library of Congress Control Number: 2014933789

Printed on acid-free paper

Springer is part of Springer Science+Business Media (www.springer.com)

To all our
Teachers, from KG to PG,
and
All Students, past and present, who are
always behind our crafted achievements!!!

—Rajesh M. Bodade and Sanjay N. Talbar

Foreword

It is a pleasure to be invited to write a preface for this book, *Iris Analysis for Biometric Recognition Systems*, by R. M. Bodade and S. N. Talbar. The human iris is well known as a stable and reliable biometric, and iris recognition is a fairly well-developed area of research now, with a few books already available on the subject. Then, what is new about this book?

The book provides a comprehensive overview of the developments in this important area of biometrics, in addition to describing some related work of the authors in the use of the complex wavelet transform.

The book has examined in detail different aspects of iris analysis, where the authors review the state of the art in the field, and present their own recent work in the area. Chapter 1 gives a nice introduction to the area. Chapter 2 presents an overview of the related research in the area. This chapter covers areas right from iris segmentation to feature extraction and recognition. It can serve as a starting point for a beginner in the area. Chapter 3 presents two simple, yet elegant, iris segmentation methods. As in some popular computer vision applications, the authors use heuristics backed by empirical studies and statistical verification. The pupil dynamics method also leads to an iris aliveness detection method.

Wavelets are not new. The use of the Dual Tree Complex Wavelet Transform (DT-CWT, hereafter), however, presents new avenues in feature extraction, which help ameliorate limitations of the traditional Discrete Wavelet Transform. The heart of the authors' recognition scheme is built around this form of wavelet transform. Texture-based features have traditionally been used for iris analysis. The authors show the use of the DT-CWT and Rotated Complex Wavelet Filters (RCWF) at different orientations and scales, at low computational cost, and provide results of extensive experimentation on standard databases in support of the proposed methodology.

The authors have created a fairly comprehensive reference for beginners and practitioners alike, in iris recognition. An important aspect of the book is its ability to explain the physical significance of many concepts in a lucid manner, which is important for an engineering practitioner, without getting lost in mathematical details. At the same time, the text provides enough pointers to a mathematically oriented reader to delve further into the myriad depths of the subject.

Sumantra Dutta Roy
IIT Delhi

Preface

Reliable personal biometric recognition is of paramount importance to modern societies, essentially due to increased use of e-commerce on one hand and rise in illegal and terrorist acts on the other hand. In this context, among other biometric traits, the iris is commonly accepted as one of the most unique and stable biometric traits and it has been successfully applied at airports and at advanced ATMs. Iris is the annular part of an eye surrounded by other unwanted parts. Therefore, fast and accurate iris segmentation and unique feature extraction are the most essential aspects of iris recognition system to achieve error-free recognition in real-time. At the same time, iris recognition system is required to be robust against counterfeit attacks.

In this book, we addressed these issues in two iris segmentation methods, one is fast and customized iris segmentation method and the other is focused on accurate methods based on pupil dynamics. Dual Tree Complex Wavelet Transform (DT-CWT) and Rotated Complex Wavelet Filters (RCWF) have been described to extract the randomly oriented multidirectional features.

Existing iris segmentation methods are either time-consuming or inaccurate and are tested on the manually edited nonrealistic CASIA database. These methods fail in segmentation of realistic images of UBIRIS database, which has been addressed in the first method using Canny operator and tangents without employing Hough transform for computational efficiency.

Most of the iris segmentation methods assume boundaries of iris as circle or ellipse, which is seldom true but exact iris is of slightly irregular shape. In the second method, we have segmented an iris of exact shape without any loss of iris data accurately by exploring the pupil dynamics of human eye. The pupil dynamics is a property of the real eye and it is also used for fake iris detection. This method has not only provided extremely high segmentation accuracy but also the excellent recognition rate, due to the loss-free, accurately segmented iris images. This novel method of iris segmentation is also inherently capable of fake iris detection.

In this book, the theories of wavelet transform and complex wavelet transform have been covered in brief. Designs of Wavelet filters, Dual Tree Complex Wavelet Transform filters, and Rotated Wavelet Filters have also been presented. These filters are used to extract randomly oriented, multidirectional texture features of an iris. We used DT-CWT and RCWF jointly to extract unique features of iris in 12 different orientations. This method successfully addressed the issue of higher computational cost of Gabor filter on one hand and inferior recognition performance of standard DWT, due to its limited directionality, on the other hand.

Thus, in this research work we designed accurate and time-efficient algorithms for all subsystems of robust iris recognition system, i.e., fake iris detection, iris segmentation, feature extraction, and matching which have been presented in five chapters. A comprehensive literature survey of existing iris recognition techniques is presented in Chap. 2. Iris segmentation methods are proposed in Chap. 3. Fast iris segmentation method customized for UBIRIS database and accurate iris segmentation algorithm using pupil dynamics are presented in this chapter. The strengths and weaknesses of our method compared to existing methods are also brought out. The inherent anti-spoofing mechanism of pupil dynamics-based iris segmentation method is outlined in this chapter. The analysis of iris texture brings out the challenges in feature extraction for iris recognition. The essence of wavelets, complex wavelets, and rotated wavelets have been explained in Chap. 4. The design of filters for complex wavelet transform and design of rotated complex wavelet filter for feature extraction are presented in this chapter. Finally, the conclusion and directions for future work are outlined in Chap. 5.

We hope this book will prove to be useful for all readers and will give future direction for postgraduate students and researchers in the area of Image Processing and Pattern Recognition. In spite of our great care, it is likely that some errors might have crept into the text. We appreciate any corrections and suggestions, which will help in the improvement of the book.

Happy reading and warm regards!!!

Indore, India   Rajesh M. Bodade
Nanded, India   Sanjay N. Talbar

Acknowledgments

It is a matter of great pride for me to have been associated with Prof. Sanjay N. Talbar, first as his research scholar and then as a coauthor of this book. It is with great pleasure that I express my deep sense of gratitude to him for his valuable guidance, constant encouragement and motivation, and support throughout this work. He has not only shaped my career as a passionate teacher, but also inspired me to write this book.

I express my gratitude to Dr. Manesh Kokare, Associate Professor, SGGS Institute of Engineering and Technology, Nanded, for his constant encouragement and strong support during the completion of this work. He has been a great source of inspiration for me to broaden my thinking in the field of pattern recognition.

I express my sincere thanks to Dr. L. M. Waghmare Director, SGGIET, Nanded, and faculty members of SGGIET, Nanded, for their valuable guidance and encouragement during my research work.

My special thanks to NASK Biometrics Laboratories, Warsaw, Poland, and research group of the laboratory for all their help, support, and guidance.

I express my sincere gratitude to the Commandant, Military College of Telecommunication Engineering (MCTE), Mhow, and the Commander, Faculty of Communication Engineering (FCE), MCTE, Mhow, for providing encouragement, all possible resources, and support to complete this book.

I express my sincere thanks to my students Jayesh Nayyar, Shailendra Ojha, Aishwairya Bhatnagar, and other students of MCTE for their help and assistance in programming and implementation of algorithms. My special thanks to Mr. Abhay Bharade for all his help and assistance in MS Office for making the manuscript of this book. I must thank my friends Manisha, Ankit, Sunil, and Kailash for fruitful discussions and constructive suggestions during the research work.

I wish to express my deepest sense of gratitude to my parents, my wife, and my daughters for their great understanding, moral support, and constant encouragement which enabled me to complete this task. My deepest thanks go to Mr. Aninda Bose of Springer for all his cooperation and support to write this book; without his trust and support it would not have been possible to complete this book. Finally, I would like to thank all those who have helped directly or indirectly in the completion of this book.

Rajesh M. Bodade

Contents

1 Introduction to Iris Recognition 1
   1.1 Introduction 1
   1.2 Iris Recognition System: Importance and Challenges 6
   1.3 Analysis of Iris for Biometric Recognition Systems 10
   References.. 11

2 Related Work... 13
   2.1 Introduction 13
      2.1.1 Daugman's Approach.. 14
      2.1.2 Wildes' Approach.. 15
   2.2 Segmentation of the Iris Region...................................... 15
   2.3 Iris Analysis and Feature Extraction 17
   2.4 Summary 20
   References.. 20

3 Iris Segmentation 25
   3.1 Introduction 25
   3.2 Iris Segmentation 27
   3.3 Fast Iris Segmentation Method Using Canny Edge
        Detector Customized for UBIRIS Database............................ 27
      3.3.1 Empirical Study of UBIRIS Database 28
      3.3.2 Outer Boundary Detection Using Canny Edge Detector 29
      3.3.3 Pupil Detection and Localization................................ 32
      3.3.4 Experimental Results 39
   3.4 Accurate Iris Segmentation Method Using Pupil Dynamics 43
      3.4.1 Flowchart of Proposed Method 45
      3.4.2 Preprocessing.. 45
      3.4.3 Outer Boundary Detection 45
      3.4.4 Pupil Detection.. 49
      3.4.5 Removal of Specular Reflections.................................. 49
      3.4.6 Normalization.. 50
      3.4.7 Fake Iris Detection 51

        3.4.8   Experimental Results   51
        3.4.9   Analysis of Experimental Results of
                Pupil Dynamics Method.   52
    3.5   Summary   55
    References.   56

**4   Iris Recognition Using Dual-Tree Complex Wavelet Transform
    and Rotated Complex Wavelet Filters**   59
    4.1   Introduction   59
    4.2   Theoretical Aspects of Wavelet Transform.   60
        4.2.1   Wavelets   61
        4.2.2   Continuous Wavelet Transforms.   61
        4.2.3   Discrete Time Wavelet Transforms   62
        4.2.4   Discrete Wavelet Transform   63
    4.3   Implementation of DWT   64
        4.3.1   Perfect Reconstruction   66
        4.3.2   Two-Dimensional Discrete Wavelet Transform   67
    4.4   Limitations of Wavelet Transforms   68
        4.4.1   Shift Sensitivity   68
        4.4.2   Poor Directionality   68
        4.4.3   Absence of Phase Information   68
        4.4.4   Aliasing   68
    4.5   Hilbert Transform and Analytic Signal   69
    4.6   Complex Wavelet Transform   70
        4.6.1   The Dual-Tree Approach for Complex Wavelets   71
        4.6.2   Selesnick's Dual Tree.   72
        4.6.3   2D DT-CWT.   73
        4.6.4   2D DT-CWT— Wavelet Filter Design   77
    4.7   DT-CWT Filters—Design and Implementation   78
        4.7.1   Design of Low-pass Filter (Scaling Function)
                of Real Tree of DT-DWT   81
        4.7.2   Design of High-Pass Filter (Wavelet Function)
                from the Low-Pass Filter   85
        4.7.3   Filter Design for Other Stages (After Stage 1)
                of DT-CWT   87
        4.7.4   Filters for 2D DT-CWT   88
        4.7.5   Rotated Complex Wavelet Filters.   88
    4.8   Experimental Results   91
        4.8.1   Role of Energy and Standard Deviation of Sub-bands
                in Iris Recognition   94
        4.8.2   Recognition Performance of Various Feature
                Extraction Methods   94
        4.8.3   Performance Analysis Using Intra-Class
                and Inter-Class Separation Test   97

4.8.4   Analysis of Size of Feature Vector and Processing Time. . . .   97
4.8.5   Shift Invariance Test of DT-CWT   100
4.9   Summary   101
References...   102

5   Conclusion and Future Scope   105

About This Book   109

Acronyms

1-D	One Dimensional
2-D	Two Dimensional
CoWT	Continuous (Analog) Wavelet Transform
CWT	Complex Wavelet Transform
DFT	Discrete Fourier Transform
DT-DWT	Dual Tree Discrete Wavelet Transform
DT-DWT(K)	Kingsbury's Dual Tree Discrete Wavelet Transform
DT-DWT(S)	Selesnick's Dual Tree Discrete Wavelet Transform
DWT	Discrete Wavelet Transform
FFT	Fast Fourier Transform
FIR	Finite Impulse Response
FT	Fourier Transform
MRA	Multi-Resolution Analysis
PR	Perfect Reconstruction
STFT	Short-Time Fourier Transform
SWT	Stationary Wavelet Transform
WP	Wavelet Packet Transform
WT	Wavelet Transform
σ	Standard Deviation

About the Authors

Rajesh M. Bodade is currently working as an Associate Professor in Faculty of Communication Engineering (FCE), Military College of Telecommunication Engineering, Mhow, Indore, India. He completed his Ph.D. in Electronics and Telecommunication Engineering (Iris Analysis for Biometric Recognition System) from SRTM University, Nanded, India in 2011. Dr. Bodade has more than 20 years of teaching experience. He received "Chief of Army Staff Medal and Commendation" and "General Officer Command in Chief-Army Training Command Medal and Commendation" on 26 January 2011 and 15 August 2013, respectively, for his outstanding contribution in teaching and research work at MCTE, Mhow. He has published 20+ papers in refereed Journals and National/ International Conferences of repute. He is Life member of Indian Society for Technical Education, Institution of Engineers (India) and Fellow of IETE. His research interests include signal processing, pattern recognition, biometrics, SDR, and cognitive radio.

Sanjay N. Talbar is a Professor in the Department of Electronics and Telecommunication Engineering, Shri Guru Gobind Singhji Institute of Engineering and Technology, Nanded, India. He received the B.E. and M.E. from SGGS Institute of Technology, Nanded, India in 1985 and 1990, respectively. He obtained his PhD from SRTM University, Nanded, India in 2000. He received the "Young Scientist Award" by URSI, Italy in 2003. He has published 10 books, 50+ papers in refereed Journals, and more than 120 papers in National and International Conferences. He has guided about 70 students for dissertation at M.E./M.Tech. and supervised 10 students for PhD and additional research scholars are working towards their PhD. He is a member of various prestigious academic committees. He is a member of IET (UK), life member of Indian Society for Technical Education (ISTE), and Fellow of IETE. His research interests include image processing, multimedia computing, biometrics, and embedded system design.

Chapter 1
Introduction to Iris Recognition

Abstract In today's networked world, the need to maintain the security of information or physical property is becoming both increasingly important and difficult. Most of the times, criminals have been taking advantage of a fundamental flaw in the conventional access control systems. The access control systems based on biometrics have a potential to overcome most of the deficiencies of current security systems and have been gaining importance in recent years. Comparison of various biometric traits shows that iris is very attractive biometric because of its uniqueness, stability, and non-intrusiveness. Number of problems required to be tackled in order to develop a successful iris recognition system, namely aliveness detection, iris segmentation, and feature extraction. However, iris is an annular part of an eye surrounded by other unwanted parts. Hence, imposes various challenges in accurate iris segmentation and feature extraction techniques to provide many opportunities for researchers in pursuing their research work in this area.

Keywords Biometric recognition system • Iris recognition • Analysis of iris for recognition system • Biometric traits • FAR • GAR • FRR • EER • ROC • Identification • Authentication

1.1 Introduction

Traditionally, we make use of passwords, PINs, and identity cards for creation of secure and reliable environment. Nowadays, in the scenario of increasing terrorism and illegal acts on one hand and increased use of e-commerce on the other hand, there is an urgent need of biometric systems that are more reliable and secure to verify a person's identity based on physical (face, fingerprint, iris, etc.) or behavioral (voice, signature, gait, etc.) characteristics because these are permanent for a person and he cannot lose or forget these characteristics in the way he loses cards or forgets passwords as in the traditional systems.

R. M. Bodade and S. N. Talbar, *Iris Analysis for Biometric Recognition Systems*,
SpringerBriefs in Computational Intelligence, DOI: 10.1007/978-81-322-1853-1_1,
© The Author(s) 2014

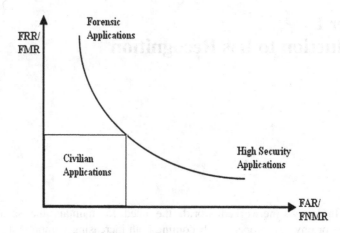

Fig. 1.1 Receiver operating characteristic (ROC) [1]

Applications of such a system include computer systems security, secure electronic banking, mobile phones, credit cards, secure access to buildings, health, and social services. Using biometrics, a person could be identified based upon 'who she/he is' rather than, 'what she/he has' (card, token, key), or 'what she/he knows' (password, PIN).

Biometric Recognition System

A biometric recognition system [1] is essentially a pattern-recognition system that recognizes a person based on a feature vector derived from a specific physiological or behavioral characteristic that the person possesses.

Biometric recognition system can operate as either *verification system* or an *identification system*. While identification involves comparing the acquired biometric information against templates corresponding to all users in the database (1 to N match) that are suitable for forensic applications such as surveillance and policing, whereas verification involves comparison with only those templates corresponding to the claimed identity (1 to 1 match) that are suitable for high-security applications like access control to highly secure places. In later case, high-security access application demands ideally zero false acceptance although false rejection is tolerable to some extent because they want to grant access to genuine persons. In forensic applications, where the desire to catch a criminal outweighs the inconvenience of examining a large number of falsely accused individuals, but false rejection shall be minimal. Most of the civil applications such as ATM access, computer login need optimum false acceptance and false rejection. The categorization of these applications is shown in Fig. 1.1 [1].

Face and voice can be considered as obvious choice for identification system (policing and surveillance applications) because of its non-intrusive acquisition, even for noncooperative subjects, without one's knowledge, whereas fingerprint and palm print can be considered better for verification system (access control and e-commerce applications) because of their better stability and uniqueness. But, due to invasiveness, image acquisition, much better stability, and highest amount of uniqueness, iris can be considered equally promising for all kind of applications.

In general, biometric trait shall possess following characteristics [2] for its suitability in recognition system.

1. *Universality*: Everyone should have it
2. *Distinctiveness*: No two should be the same
3. *Permanence*: It should be invariant over a given period of time
4. *Collectability*: It should be measurable by easy means
5. *Circumvention*: It should be robust enough to various fraudulent methods
6. *Acceptability*: It must be harmless to users
7. *Performance*: It is related to accuracy, speed, and resource requirements for its implementation.

The comparison of various biometric traits with respect to these characteristics is given in Table 1.1 [2].

Figure 1.2 shows the Zephyr analysis of various biometric traits with reference to their variations in intrusiveness, accuracy, cost, and ease of sensing [3, 4]. Thus, it is clear that no single biometric can be considered as the 'best' recognition tool and the choice depends on the application. However, iris seems to be optimum choice in recognition system.

Biometric System Performance Measures

The performance of biometric recognition systems is measured with the help of following parameters [5].

False Acceptance Rate (FAR) or False Match Rate (FMR)

It measures the probability of confusing two identities, and it is the rate of accepting false user. Obviously, this is the most important measure, regarding security. FAR can be defined as:

$$FAR = \frac{\text{Number of times different person matched} \times 100}{\text{Number of comparison between different persons}} \qquad (1.1)$$

False Reject Rate (FRR) or False Non-Match Rate (FNMR)

It has great relevance in the comfort that the biometric system affords to its users and it is the rate of rejecting genuine user. FRR can be defined as:

$$FRR = \frac{\text{Number of times same person rejected} \times 100}{\text{Number of comparison between same persons}} \qquad (1.2)$$

Genuine Accept Rate (GAR) or True Accept Rate (TAR)

It is the alternative measure for FRR, and it can be defined as:

$$TAR = GAR = 1 - FRR \qquad (1.3)$$

Table 1.1 Comparison of various biometric traits [2]

Biometric trait	Biometric characteristic						
	Universality	Distinctiveness	Permanence	Collectability	Performance	Acceptability	Circumvention
Hand vein	M	M	M	M	M	M	L
Gait	M	L	L	H	L	H	M
Ear	M	M	H	M	M	H	M
Fingerprint	M	H	H	M	H	M	M
Face	H	L	M	H	L	H	H
Retina	H	H	M	L	H	L	L
Iris	*H*	*H*	*H*	*M*	*H*	*L*	*L*
Voice	M	L	L	M	L	H	H
Signature	L	L	L	H	L	H	H
DNA	H	H	H	L	H	L	L

L Low, *M* Medium, *H* High

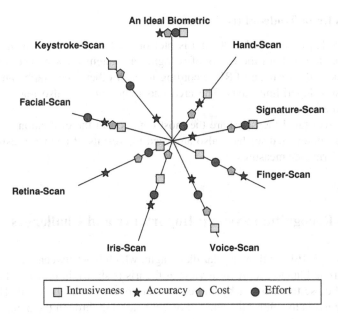

Fig. 1.2 Zephyr analysis of biometric traits [3]

Recognition Accuracy

The overall recognition accuracy is defined as:

$$\text{Accuracy} = 100 - \frac{\text{FRR} + \text{FAR}}{2} \tag{1.4}$$

Equal Error Rate (EER)

It is a very common measure of the biometric systems accuracy. This is the value where both errors (FAR and FRR) become approximately equal.

FAR and FRR are dual measures and meaningful exclusively when presented together. FAR and FRR are functions of the system threshold. As threshold decreases, FAR increases and FRR decreases and vice versa. The trade-off between FAR and FRR depends on the security and throughput requirements of the system and are highly dependent upon the specifications of the application environment.

Receiver Operating Characteristic (ROC)

ROC [5] is plot of FAR on *x*-axis and TAR on *y*-axis, which reflects the variability of both the FAR and GAR in accordance with each other. ROC curves are threshold independent, allowing performance comparison of different systems under similar conditions, or of a single system under differing conditions.

Detection Error Trade-off (DET) Curve

DET curve [5] is modified ROC and is plot of FAR versus FRR which provides the information of both the errors of recognition system. This reflects the variability of both the FAR and FRR according to each other. The graph can then be plotted using logarithmic axis. The area under the curve is also regarded as an accuracy measure.

The above-stated measures are the most common in the evaluation of the recognition accuracy, and we have also analyzed the results of our work using these above performance measures.

1.2 Iris Recognition System: Importance and Challenges

The human iris [6] is a thin circular diaphragm, which lies between the cornea and the lens of the human eye. A front view of the iris is shown in Fig. 1.3. The iris is perforated close to its center by a circular aperture known as the pupil. The function of the iris is to control the amount of light entering through the pupil, and this is done by the sphincter and the dilator muscles, which adjust the size of the pupil. The average diameter of the iris is 12 mm, and the pupil size can vary from 10 to 80 % of the iris diameter [7].

The iris consists of a number of layers, the lowest is the epithelium layer, which contains dense pigmentation cells. The stromal layer lies above the epithelium layer and contains blood vessels, pigment cells, and the two iris muscles. The density of stromal pigmentation determines the color of the iris. The externally visible surface of the multilayered iris contains two zones of different colors [8]. An outer zone is called sclera zone, and the inner one is pupillary zone. These two zones are divided by an iris that is made up of the collarette, which appears in random pattern.

Formation of the iris begins during the third month of embryonic life. The unique pattern on the surface of the iris is formed during the first year of life, and pigmentation of the stroma takes place for the first few years. Formation of the unique patterns of the iris is random and is not related to any genetic factors [9].

The only characteristic that is dependent on genetics is the pigmentation of the iris, which determines its color. Due to the epigenetic nature of iris patterns, the eyes of an individual contain completely independent iris patterns, and identical twins possess uncorrelated iris patterns. Iris is the ring [9] of colored tissues between white sclera and dark pupil of an eye as shown in Fig. 1.3. Light enters inside the eye to reach the retina through the pupil. The size of the iris varies to adjust the amount of light entering the pupil. The iris typically has a rich pattern of furrows, ridges, and pigment spots. The color of the iris can change as the amount of pigment in the iris increases during childhood. The minute details of iris texture are believed to be random, unique, and very stable throughout lifetime of a person.

Although, the idea of using iris for recognition is very old, John Dougman, in 1993, described the working iris recognition system [10, 11]. It consists of image

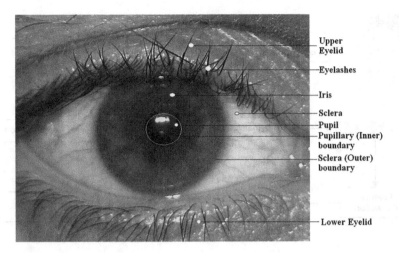

Fig. 1.3 Human eye and its parts

acquisition unit, iris segmentation technique, iris analysis, and feature extraction technique and matching of extracted feature of iris.

The strength of traditional system of person verification lies in uniqueness, and stability of passwords and cards, which are the primary pillars of any verification or an identification system. In this context, from the analysis of Table 1.1, among all biometrics, iris pattern gains increasing attention for its stability, reliability, uniqueness, noninvasiveness, and difficulty in counterfeiting. Moreover, most of the currently deployed commercial algorithms [6, 10–13] for iris recognition (by John Daugman) have a very low false acceptance rate compared with the other biometric identifiers, and it is acknowledged that iris recognition is more accurate than any other biometric technique. Hence, it seems to be best suited to replace traditional system of person verification and identification as compared with other biometric traits.

However, it has many challenges in terms of image acquisition of non-cooperative subjects, accurate segmentation of iris part from all eye images, iris analysis, and its appropriate representation to create distinct feature vector to achieve high accuracy of recognition and robust anti-spoofing mechanism to detect all possible types of fake irises. Therefore, many researchers have shown lot of interest in iris biometrics in recent years.

A block diagram of robust iris recognition system is shown in Fig. 1.4. It involves five main modules, namely, image acquisition, iris aliveness detection, iris segmentation, normalization, feature extraction, and matching.

Image Acquisition

A specially designed sensor is used to capture a sequence of iris images of the subject. Capturing good quality iris images acceptable for practical applications is one of the major challenge because the iris is comparatively small part of human face

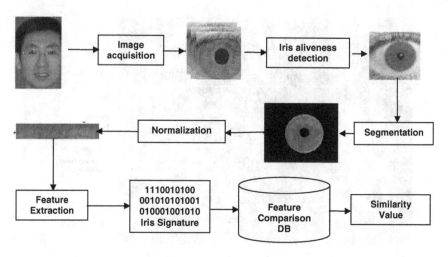

Fig. 1.4 Typical stages of the robust iris recognition systems [14]

surrounded by eyelids and eyelashes, and it exhibits more abundant texture features under infrared lighting [14]. Fortunately, much work has been done on iris image acquisition, which has made noninvasive imaging at a distance possible [15, 16].

Iris Aliveness Detection

Iris aliveness detection aims to ensure that an input image sequence is from a live subject instead of an iris photograph, a video playback, a glass eye, etc. Although, iris is believed to allow very high accuracy and it is difficult to be forged as compared with other popular biometric traits such as fingerprint and face, various experiments showed an alarming lack of anti-spoofing mechanisms in devices already protecting many sensitive areas all over the world [17]. This enforces the need for aliveness detection methodology to be quickly introduced and stressed the need for the research on fake resistive iris recognition. Daugman [18, 19] introduced the method of using Fast Fourier Transform in order to check the printed iris pattern and also mentioned without much details the reflectance method for liveliness detection of irises at very preliminary level.

Although, iris liveliness detection is highly desirable, very limited practical efforts by few research groups [18–22] are done based on active and passive methods separately to detect few types of fake irises. The utilization of the optical and physiological characteristics of the live eye to implement effective aliveness detection of all possible types of fake irises is a challenge and remains to be an important research topic.

Iris Segmentation

An eye image, as shown in Fig. 1.3, contains not only the region of interest (iris) but also noise in the form of pupil, sclera, eyelids, and eyelashes. The iris ring should be extracted from the captured image. To extract the iris ring, the inner boundary between the pupil, the iris, and the outer boundary of the iris has to be detected with help of suitable algorithm. Daugman's original approach [10–13] of iris segmentation uses an integro-differential operator, and Wildes [9] suggested a method involving edge detection and a Hough transform. The major challenges in accurate iris segmentation of images with occlusion of iris by eyelids and eyelashes are as follows: off-angle images, specular reflections in images, low-intensity gradient at inner and outer iris boundary of high-resolution images captured in less constrained imaging environments, either under natural luminosity, from different image capturing distances and without user's cooperation. Variants of the stated approaches have since been used by a number of researchers with certain modifications in order to improve the performance by overcoming the above-stated (one or multiple) challenges.

Iris Normalization

Segmented iris ring from an eye image has various inconsistencies. The sources of inconsistencies include stretching of the iris caused by pupil dilation from varying levels of illumination, varying imaging distance, rotation of the camera, head tilt, and rotation of the eye within the eye socket, etc. In order to compensate these inconsistencies, it is transformed from Cartesian system of coordinates to dimensionless polar system of coordinates. This is nothing but a normalization process. This is usually accomplished through a method similar to the Daugman's rubber sheet model [10–13].

Feature Extraction

Since the resolution of an image can be very high, it is not practical in efficiency, storage and accuracy to take every pixel of the image as a feature and describe it by a vector. It is necessary to extract only a subset of pixels from an image, either in original domain or in transformed domain, to be described. This subset of pixels is called the interest points or feature vector. There are two main requirements on feature detector. First, corresponding interest points on the object should be repeatedly detected by the feature detector over different images of the same object. Second, interest points detected should carry distinctive information.

In order to provide accurate recognition of individuals, the limited but most discriminating information present in an iris is extracted and feature vector of significant features of the iris is formed. These features are either local or global features of iris extracted either in spatial domain or in transformed domain or can be in the statistical form.

Daugman's approach is based on Gabor filters to produce a binary representation of iris called as iris code [10–13], and Wildes' approach [9, 15, 16] is based on Laplacian of Gaussian filters at multiple scales to represent iris with a feature vector. These two main approaches in iris recognition motivated researchers to investigate various methods of feature extractions that are broadly categorized into three groups, namely, filter-based methods, transform-based method, and texture-based statistical methods.

The extracted features are stored in database. The iris image to be tested is converted into feature vector, and then, it is compared with database feature vector for recognition purpose. The simple approach is template matching in iris recognition, where all iris images are stored in database and input image will be compared one by one for matching or recognition.

Looking at different approaches, analyzing the texture of the iris has perhaps been the most popular area of research in iris biometrics. The main challenge in feature extraction is to represent randomly oriented texture of iris that is available in various directions and orientations with the help of computationally efficient method to produce compact feature vector so that improved recognition rate at reduced processing time and reduced computational cost is achieved.

1.3 Analysis of Iris for Biometric Recognition Systems

From the literature survey on iris recognition, following points related to iris segmentation, feature extraction, and fake iris detection are revealed.

1. Iris segmentation has been carried out mostly assuming circular papillary and limbic boundaries that are seldom true. Although, few researchers assumed elliptical shape for iris segmentation to achieve better results, accurate iris segmentation without assumption of any fixed shape has not been addressed.
2. Variation in pupil size due to change in illumination is termed as pupil dynamics and has been used for fake iris detection. However, use of pupil dynamics for iris segmentation has not been researched.
3. Iris is a rich texture image, containing frequency and orientation information in multiple directions. Most of the methods used either one or other kind of wavelet transform for computational efficiency or Gabor filter for appropriate iris representation to extract the iris features. Each has its advantages and limitations. However, DT-CWT and RCWF can provide the features in more orientations at low-computational cost as compared with Gabor filters. But, use of these techniques has not been researched for iris recognition so far.

In view of above-stated inferences, following research objectives have been addressed and discussed in this book.

1. Fast iris segmentation method customized for images of realistic UBIRIS database.
2. A novel method for accurate iris segmentation based on pupil dynamics to address challenges related to iris segmentation. This method is also capable of fake iris detection.
3. Design of the suitable filters for 2-D DT-CWT and RCWF for iris feature extraction in multiple directions and at various scales and the iris recognition based on combination of DT-CWT and RCWF to address the strengths and weaknesses of Gabor filters and DWTs.

This book consists of five chapters. Comprehensive literature survey of existing iris recognition techniques is presented in Chap. 2. Iris segmentation methods have been proposed in Chap. 3. Fast iris segmentation method customized for UBIRIS database and accurate iris segmentation algorithm using pupil dynamics are presented in this chapter. The strengths and weaknesses of our method compared with existing methods are also brought out. The inherent anti-spoofing mechanism, pupil dynamics-based segmentation method to detect active fake irises is also pointed out in this chapter. The analysis of iris texture brings out the challenges in feature extraction for iris recognition. Essence of wavelets, complex wavelets, and rotated wavelets has been explained in Chap. 4. The design of filters for complex wavelet transform and design of rotated complex wavelet filter for iris feature extraction is presented in this chapter. Finally, conclusion and directions for future work are outlined in Chap. 5.

References

1. S. Prabhakar, S. Pankanti, A.K. Jain, Biometric recognition: Security and privacy concerns. IEEE Secur. Priv. Mag. 33–42 (2003)
2. A. Jain, A. Ross, S. Prabhakar, An introduction to biometric recognition. IEEE Trans. Circuits Syst. Video Technol. **14**(1), 4–19 (2004)
3. X. Lu, Image analysis for face recognition. Available on http://www.cse.msu.edu/~lvxiaogu/publications/
4. R. Hietmeyer, Biometric identification promises fast and secure processing of airline passengers. J Int Civil Aviat Organ **55**(9), 10–11 (2000)
5. A. Mansfield, J. Wayman, Best practices in testing and reporting performance of biometric devices. Version 2.01, (2002). PDF available on: http://www.npl.co.uk/upload/pdf/biometrics_bestprac_v2_1.pdf
6. L. Masek, Recognition of human iris patterns for biometric identification. Master's thesis, University of Western Australia, (2003). Available on: http://www.csse.uwa.edu.au/~pk/studentprojects/libor/LiborMasekThesis.pdf
7. J. Daugman, How iris recognition works, In *Proceedings. of International Conference. on Image Processing,* vol. 1 (2002), pp 33–36
8. E. Wolff, *Anatomy of the Eye and Orbit*, 7th edn. (H. K. Lewis & Co, Ltd, 1976)
9. R. Wildes, Iris recognition: an emerging biometric technology. Proc. IEEE **85**(9), 1348–1363 (1997)
10. J. Daugman, High confidence visual recognition of persons by a test of statistical independence. IEEE Trans. Pattern Anal. Mach. Intell. **15**(11), 1148–1161 (1991)
11. J. Daugman, Biometric personal identification system based on iris analysis. U.S. Patent No. 5,291,560, (1994)
12. J. Daugman, The importance of being random: statistical principles of Iris recognition. Pattern Recogn. **36**(2), 279–291 (2003)
13. J. Daugman, How iris recognition works. IEEE Trans. Circuits Syst. Video Technol. **14**(1), 21–30 (2004)
14. L. Ma, T. Tan, Y. Wang, D. Zhang, Personal identification based on iris texture analysis. IEEE Trans. Pattern Anal. Mach. Intell. **25**(12), 1519–1533 (2003)
15. R. Wildes, J. Asmuth, S. Hsu, R. Kolczynski, J. Matey, S. Mcbride, Automated noninvasive iris recognition system and method. United States Patent, No. 5572596, (1996)
16. R. Wildes, J. Asmuth, G. Green, S. Hsu, R. Kolczynski, J. Matey, S. McBride, A machine-vision system for iris recognition. Mach. Vis. Appl. **9**, 1–8 (1996)

17. L. Thalheim, J. Krissler, P. Ziegler, Biometric access protection devices and their programs tut to the test (2002), p. 114. Available on http://www.larc.usp.br/~pbarreto/Leitura%202%20 -%20Biometria.pdf
18. J. Daugman, Iris recognition and anti-spoofing countermeasures, In *Proceedings. of 7th International Conference on Biometrics,* London, (2004)
19. J. Daugman, Recognizing persons by their iris patterns: countermeasures against subterfuge, in *Biometrics* ed. by Jain et al. (Personal Identification in a Networked Society, 1999), pp. 103–121
20. S. Lee, K. Park, J. Kim, A study on fake iris detection based on the reflectance of the iris to the sclera for iris recognition, in *Proceedings of ITC–CSCC Conference-2005,* Jeju Island, South Korea, pp. 1555–1556 (2005)
21. A. Pacut, A. Czajka, Aliveness Detection for Iris Biometrics, in *Proceedings. of IEEE International Carnahan Conference. on Security Technology,* pp. 122–129 (2006)
22. M. Clynes, M. Cohn, Color dynamics of the pupil. Ann. N.Y. Acad. Sci. **156**, pp. 931–950 (1969) (*John Wiely Online Library,* 2006)

Chapter 2
Related Work

Abstract Compressive literature survey has been carried out and essence of it has been presented in this chapter. It reveals that in 1987, Flom and Safir proposed the first conceptual but unimplemented automated model of Iris recognition system. In 1992, Johnson analyzed Iris images and confirmed its high stability over a period of 15 years. Based on Flom and Safir model, Daugman in 1993 and Wildes in 1997 had proposed two complementary approaches of Iris recognition system and most of the research in this field is motivated and based on either of the two approaches. Related work carried out in iris segmentation, iris analysis, and feature extraction in last two decades has been presented and analyzed in this chapter. Either of the approaches, namely binary representation of Iris or real valued feature vector of Iris, has been explored very extensively by many researchers, mainly, either by using variants of Gabor filters or by using DWT for multi-resolution representation of Iris. Various iris image databases used by various research groups are also studied, and it is observed that CASIA database, which is less realistic has been explored more than realistic databases such as UBIRIS, UPOL.

Keywords Iris analysis and feature extraction • Daugman's approach • Wildes' approach • Flom and safir model • Gabor filters • Laplacian of Gaussian (LOG) filter • Iris code

2.1 Introduction

In 1987, Flom and Safir proposed the first conceptual but unimplemented automated model of Iris recognition system [1]. They suggested highly controlled and non-practical conditions to change the illumination so that pupil size in all images remains same for proper Iris segmentation. They outlined the major subsystems of Iris recognition system, namely image acquisition unit, preprocessing and Iris segmentation unit, Iris analysis and feature extraction unit, and matching unit along

R. M. Bodade and S. N. Talbar, *Iris Analysis for Biometric Recognition Systems*, SpringerBriefs in Computational Intelligence, DOI: 10.1007/978-81-322-1853-1_2, © The Author(s) 2014

with suitable image processing and pattern recognition methods. This theoretical work on Iris recognition has proved as a foundation for all practical approaches of Iris recognition.

In 1992, before Daugman's work, Johnson [2] analyzed 650 Iris images in two sessions with a gap of 15 years and observed no change in the pattern of Irises. Thus, Iris is one of the highly suitable biometric traits for both person authentication and person identification due to its high stability.

Based on Flom and Safir model, Daugman in 1993 and Wildes in 1997 had proposed two complementary approaches of Iris recognition system, and most of the research in this field is motivated and based on either Daugman's approach or on Wildes' approach. The outline of these two main approaches has been stated below.

2.1.1 Daugman's Approach

In 1993 [3], John Daugman proposed most relevant system, forming the basis for all operational systems. His patent [4], in 1994, described the first functional Iris recognition system. His publications [5–7] stated the use of near infrared (NIR) illumination for image acquisition. NIR illumination not only remains nonintrusive to humans but helps to reveal the detailed structure of heavily pigmented Irises also.

For Iris segmentation, he assumed inner and outer boundaries of an eye as circles [3] which are characterized by radius, r, and coordinates of circle, x_0 and y_0. He introduced the integro-differential operator for detecting both inner and outer boundaries which is given by Eq. (2.1).

$$\max(r, x_0, y_0) \left| G_\sigma(r) * \frac{\partial}{\partial r} \oint_{r,x_0,y_0} \frac{I(x,y)}{2\pi r} ds \right| \qquad (2.1)$$

where $I(x,y)$ is the eye image and $G_\sigma(r)$ is Gaussian filter for smoothening.

This method tries to find a circle in the image with maximum gray level differences with its neighbors. Inner circle is determined first due to higher intensity gradient at inner boundary. Then, outer boundary is detected using same operator by searching the new parameter space. Then, the segmented Iris image is converted to dimensionless polar coordinate system for size normalization to compensate for the imaging inconsistencies.

Daugman used 2-D complex-valued Gabor filter to extract texture phase information of Iris, and each complex phase coefficient is quantized and encoded to obtain 256 byte binary 'Iris code' [3]. The compact and convincing representation of Iris with the help of binary code of 2-bit quantized phase information is the distinctive feature of Daugman method. Similarity between a pair of Iris code is measured by their normalized Hamming distance based on exclusive-OR operation.

2.1.2 Wildes' Approach

Wildes developed an Iris recognition system at Sarnoff Laboratory [8] with different approach as compared to Daugman's approach. He used a low-light-level camera along with a diffused source and polarization, for image acquisition.

He proposed the most common method of Iris segmentation through a gradient-based binary edge-map construction followed by circular Hough transform (CHT). Issues of eyelid noise and its removal also have been addressed in the method. This method is more stable to noise, but due to binary edge abstraction, it losses some data of Iris, which can be crucial in feature extraction.

Wildes used Laplacian of Gaussian filter at multiple scales to create a feature template. This template is a lesser compact representation of Iris because it incorporates finer distinctive data which is used to compute normalized special correlation as a similarity measure at matching stage. Wildes tested his method on several hundred Iris images. Wildes et al. [9] also filed patents for normalized spatial correlation for matching and user friendly image acquisition setup.

2.2 Segmentation of the Iris Region

This is one of the very important stages of Iris recognition system, and many researchers proposed various approaches. This stage consists of detection of pupillary (outer) and limbic (inner) boundaries, removal of pupil and sclera, detection, and removal of eyelids and eyelashes. As stated above, Daugman used integro-differential operator, whereas Wildes et al. used edge detection and CHT for Iris segmentation.

Ma et al. [10–12] has carried out the detailed analysis of Iris texture information and reported that Iris region closer to pupil provides the most useful texture information for recognition [10]. They approximated inner and outer boundary as circles and segmented the Iris using approach similar to Wildes' approach in two steps. After normalization, an image enhancement by subtracting an average of Iris from normalized Iris has been proposed in their work. Moreover, eyelashes and eyelids rarely occlude this region.

Most of the work in segmentation is based on Wildes et al. approach to improve one or the other aspect. Huang et al. [13] proposed variant of Wildes' method to reduce the computational complexity by finding the Iris boundaries in rescaled image in first step and then using that information to guide search on original image in next step. They also proposed the idea of making Iris rotation invariant for matching by making use of image of other eye as reference.

Similarly, use of Canny edge detector and Hough transform in a simplified manner is described by Liu et al. [14]. They proposed filtering of Iris normal region by using low-frequency filter. In addition to the detection of Iris boundaries, collarette boundary detection is proposed by Sung et al. [15] to increase the recognition rate. For finding the collarette boundary, histogram equalization and a high-pass filter, after using a one-dimensional DFT, are applied to the image. The collarette boundary is found using statistical information from the image.

Iris segmentation algorithm based on texture segmentation is proposed by Cui et al. [16] by using low frequency of wavelet transform of the Iris image for pupil segmentation and localize the Iris with a differential integral operator. The lower eyelid is localized using parabolic curve fitting based on gray value segmentation.

Kong and Zhang [17] used edge detector and CHT on nonlinearly enhanced eye image for Iris segmentation in first step, and in next step, eyelashes and eyelid interference has been removed from an Iris image. Lili and Mei [18] adopted edge point detection and curve fitting for Iris segmentation, and Iris image quality test has been conducted to discard poor quality Irises.

Teo and Ewe [19] utilized black hole search method to detect pupil because pupil is the darkest region in the image. This method is not suitable for eye image with dark Iris. Similar approach is proposed by Grabowski et al. [20].

He and Shi [21] carried out the binarization of image to detect the pupil, and then, with the help of edge detection and Hough transform, outer boundary is detected. Feng et al. [22] used a 'course-to-fine' approach to find the inner and outer boundaries of Iris by approximating them as circles. One of the important findings of their research is that use of the lower contour of the pupil in obtaining the inner boundary because it is stable even in the seriously occluded image. Tian et al. [23] proposed the method of searching a pixel of low intensity to detect an approximate pupil center and then use edge detection and Hough transform for boundary detection and Iris segmentation.

Prior pupil segmentation approach was proposed by Du et al. [24]. Further, they used polar coordinates and Sobel operator to detect the outer boundary by assuming concentric circles of Iris and pupil.

Camus and Wildes [25] proposed a method which was somewhat similar to Daugman's approach. This method was based on N^3 space search of three parameters: x, y, and r. The parameters were tuned to maximize the goodness-of-fit criteria to obtain accurate Iris segmentation. Roche et al. [26] also utilized N^3 space search for three circumference parameters (center (x, y) and radius r) where the difference between the average intensity of five successive circumferences is maximal. They made use of histogram stretching to maximize average intensity differences of the circumferences.

Most of the researchers used CASIA version 1 [27] image database for their research, but Phillips et al. [28] had reported that the pupil area in each image of this database had been replaced with the circular region of constant intensity (black) to mask out the specular reflections from the near infrared illumination source. This intestinally edited image database certainly reduces the challenges in Iris segmentation and calls into question any results obtained using it as it has made Iris segmentation artificially easy.

Therefore, Proenca et al. [29] evaluated clustering algorithms for preprocessing the images to enhance image contrast, and they tested not only their method on the

UBIRIS dataset [30] but also tested the methods of Daugman [3], Wildes [8], and other methods, which contains one session of high-quality images, and a second session of lower quality images. The comparative analysis found that the results of all methods other than Wildes method have underperformed compared to their original results.

Unlike many others, Bonney et al. [31] modeled inner and outer boundaries as ellipses instead of circles. They made use of least significant bit plane to detect pupil and then obtain the elliptical limbic boundary by computing the standard deviation in the horizontal and vertical directions.

Iris segmentation of off-angled images has been carried out by estimating the elliptical shape rather than circular shape by Li [32] and Abhyankar et al. [33]. Moreover, Abhyankar et al. [33] described that an elliptical shape boundaries of Iris can represent Iris more accurately than circular boundaries and in [34]; they used active shape models to detect the non-circular (elliptical) boundaries.

Detailed study of images of CASIA [27], UBIRIS [30], and UPOL [35] databases shows that change in the intensity (intensity gradient) at outer and inner boundary of Iris is maximum in CASIA database, moderate in UBIRIS database, and minimum in UPOL database.

2.3 Iris Analysis and Feature Extraction

As stated earlier, Daugman's approach is based on Gabor filters to produce a binary representation of Iris called as Iris code and Wildes' approach is based on Laplacian of Gaussian filters at multiple scales to represent Iris with a feature vector. These two main approaches in Iris recognition-motivated researchers to investigate various methods of feature extractions which are broadly categorized into three groups, namely filter-based methods, transform-based method, and texture-based statistical methods.

Modified Log-Gabor filters are used by Yao et al. [36] instead of Gabor filters for feature extraction because LOG-Gabor filters are strictly band-pass filters compared to Gabor filter. They state that Gabor filters could not represent high-frequency components in natural images properly. Hence, using the modified filters, EER is improved marginally. Zhang et al. [37] also used LOG-Gabor filter to find local and global texture feature because zero DC component can be obtained for any bandwidth by using a LOG-Gabor filter. They stated normal Iris code as local features and global features are considered to be invariant to Iris rotation and minor errors in Iris segmentation. They proposed a cascaded system, first based on global features and second on local features.

Sun et al. [38] proposed local ordinal encoding using of Gaussian filter with the gradient vector field of an Iris image. They also proposed a general framework of Iris recognition which is useful in Iris representation and is formulated in this paper. Sun et al. [39] also proposed a cascaded system of two stages, first stage is Daugman-like approach and second stage looks at global features, i.e., 'areas enclosed by zero-crossing boundaries.'

Park et al. [40] used directional filter bank for Iris decomposition and computed two feature vectors, one as binarized directional sub-band outputs and other as block-wise directional energy values. They use these feature vectors independently for recognition, and final result is declared after combination. Improvement has been reported due to combination.

Similar to Wildes approach, Chenhong et al. [41] used Laplacian of Gaussian filters to compute the discriminable textons from geometric and luminance attributes of texture elements. In this method, morphological operations with two threshold are also employed to segment Iris on the basis of local shape into small compact and thin elongated components. Chou et al. [42] used both derivative of Gaussian and Laplacian of Gaussian filters to find whether a pixel is on step edge or ridge edge. Major motivation for use these types of filters is that only three filter parameters are used for filter design, and hence, they can be easily determined.

Ma et al. [10] described the modification of elementary Gabor filter to circularly symmetric Gabor filters which are suitable to capture local details of Iris, and they used it at two scales to obtain the prominent information in many directions and more prominently in x and y directions. They stated that, 'the performance of their system reached very close to Daugman's system and all other methods [8, 43].' They concluded that phase information characterized by Daugman method provides local shape features of an Iris, whereas their method provides statistical information of the frequency information of a local region.

Boles et al. [43] converted concentric bands of extracted Iris into 1-D signals and operated wavelet transform on it to generate the zero-crossing presentation. Energy of zero-crossing representation is used as one feature, and dimensions of rectangular pulses of zero-crossing representations are used as other feature.

Similar approach is also presented by Sanchez-Avila et al. [44] to compare the Gabor filter-based Daugman approach with dyadic wavelet transform-based zero-crossing approach and reported that Gabor filter approach achieves better performance than wavelet transform-based approach but wavelet transform-based approach showed better computational efficiency over Gabor filter-based approach. Krichen et al. [45] used wavelet packets for the feature extraction of images captured at visible light.

Thornton et al. [46] has considered seven different types of filters including Gabor wavelet for comparison and determined that Gabor wavelet gave the best recognition results among all. However, they brought out an interesting fact that the performance of Gabor wavelet is highly dependent upon parameters that define its form, and proper tuning of parameters is required for its optimum performance.

Kim et al. [47] used disk-shaped Iris image without normalization, and it is first convolved with a low-pass filter along the radial direction. Then, one-dimensional wavelet transform is operated on the smoothened Iris image to obtain angular directions of decomposed images which is approximated by an optimal piecewise linear curve connecting a small set of node points. The set of node points is used as a feature vector.

Tisse et al. [48] represented an Iris with the help of instantaneous phase. The *instantaneous* phase is obtained by constructing the *analytic signal*, which is the combination of the original signal and its *Hilbert* Transform. Miyazawa et al. [49]

has computed discrete Fourier transform (DFT) of Iris image and used phase information to create the feature vector for correlation.

Hosseini et al. [50] proposed a shape analysis approach for Iris recognition using FFT-based Tikhonov filter with the identity matrix as the regularization operator to represent shape of pigmented fibro-vascular tissue known as stroma which is extracted by operating adaptive filter.

Recently, in 2009, Azizi and Pourreza [51] have proposed contourlet transform-based method of feature extraction method for Iris recognition. Contourlet transform captures the intrinsic geometrical structures of Iris image to decompose the Iris image into a set of directional sub-bands with texture details captured in multiple orientations at multiple scales.

Statistical methods are also used for Iris analysis and feature extraction by limited researchers. As compared to Iris recognition, substantial research has been carried out in face recognition using PCA and ICA [52, 53]. Moreover, PCA and ICA have been found to be very successful methods of feature extraction for face recognition. Son et al. [54] used DWT, principal component analysis (PCA), linear discriminant analysis (LDA), and direct linear discriminant analysis (DLDA) for feature extraction, and experimented these features in combinations for Iris recognition. They concluded that DWT is best for feature extraction and DLDA is better to reduce the dimensionality of feature vector. Marian Stewart Bartlett et al. [55] used ICA with NN and presented results on frontal faces. Ekenel and Sankur [56] proposed the methodology of selection of number of independent components for feature selection and extraction using ICA.

In 2002, Huang et al. [13] experimented independent component analysis to extract the independent components of normalized Iris image using global approach. Similar approach of global processing using ICA has been presented by Dorairaj et al. [57]. However, Bae et al. [58] used ICA using local processing approach to extract the features of sub-images of Iris.

In 2008, Bowyer et al. [59] presented a systematic and detailed survey paper on Iris recognition which has covered existing research of all essential elements of Iris recognition system. This paper has been instrumental systematic study and provided comparison of various research work in Iris recognition which has greatly simplified our work of literature survey to proposed new methods of Iris segmentation and Iris feature extraction.

Selesnick et al. [60], Selesnick [61], and Nick Kingsbury [62] proposed the practical implementation of complex wavelet transform (CWT) using dual tree discrete wavelet transform (DT-DWT), in which one tree of DWT acts as real tree and other acts as imaginary tree. The low-pass and high-pass filters of both the trees are designed in such a way that second tree forms an approximate Hilbert transform of first tree. Oriented non-separable 2-D complex wavelet transform is derived by combining the sub-bands of two separable 2-D DWTs [60–62]. This transform is capable of providing orientations in six directions, phase information, and it is shift invariant too.

Kim and Udpa [63] designed new 2-D non-separable and oriented rotated discrete wavelet filter (RDWF) by rotating 2-D DWT filters by 45° for clear characterization of diagonally oriented texture. These filters clearly provide edges in the direction of 45° and −45° separately.

Kokare et al. [64] extended the work of Kim and Udpa [64] to derive the 2-D non-separable filters by rotating 2-D non-separable CWT filters, called as rotated complex wavelet filters (RCWF) to obtain the orientations in another six directions. They used these filters in combination with CWT for content-based image retrieval and reported the extremely good improvement in results as compared to DWT and CWT.

2.4 Summary

An attempt has been made to highlight the important findings of different researchers in the area of Iris recognition over the last two decades with critical observations about merits and demerits of different techniques adopted. The outcomes of this literature survey has been analyzed to define the research goals, and these outcomes have been explored to devise new techniques of Iris segmentation and feature extraction in chapter three and four, respectively.

References

1. L. Flom, A. Safir, Iris recognition system, U.S. Patent 4,641,349, 1987
2. R. Johnston, Can iris patterns be used to identify people? *Los Alamos National Laboratory, Chemical and Laser Sciences Division Annual Report LA-12331-PR*, pp. 81–86 (1992)
3. J. Daugman, High confidence visual recognition of persons by a test of statistical independence. IEEE. Trans. Pattern. Anal. Mach. Intell. **15**(11), 1148–1161 (1993)
4. J. Daugman, Biometric personal identification system based on iris analysis. U.S. Patent No. 5,291,560, 1994
5. J. Daugman, High confidence visual recognition of persons by a test of statistical independence. IEEE Trans. Pattern Anal. Mach. Intell. **15**(11), 1148–1161 (1991)
6. J. Daugman, The importance of being random: statistical principles of iris recognition, Pattern Recogn. 279–291 (2003)
7. J. Daugman, How iris recognition works. IEEE Trans. Circ. Syst. Video Technol. **14**(1), 21–30 (2004)
8. R. Wildes, Iris recognition: an emerging biometric technology. Proc. IEEE **85**(9), 1348–1363 (1997)
9. R. Wildes, J. Asmuth, S. Hsu, R. Kolczynski, J. Matey, S. Mcbride, Automated noninvasive iris recognition system and method, United States Patent, no. 5572596, (1996)
10. L. Ma, T. Tan, Y. Wang, D. Zhang, Personal identification based on iris texture analysis. IEEE Trans. Pattern Anal. Mach. Intell. **25**(12), 1519–1533 (2003)
11. L. Ma, Y. Wang, T. Tan, Iris recognition using circular symmetric filters. in *Proceedings of the 25th International Conference on Pattern Recognition (ICPR02)*, vol. 2, pp 414–417 (2002)
12. L. Ma, Y. Wang, D. Zhang, Efficient iris recognition by characterizing key local variations. IEEE Trans. Image Process. **13**(6), 739–750 (2004)
13. Y. Huang, S. Luo, E. Chen, An efficient iris recognition system, in *Proceedings of International Conference on Machine Learning and Cybernetics*, vol. 1, pp. 450–454 (2002)

14. Y. Liu, S. Yuan, X. Zhu, Q. Cui, A practical iris acquisition system and a fast edges locating algorithm in iris recognition, in *Proceedings of IEEE Conference on Instrumentation and Measurement Technology*, pp. 166–168 (2003)
15. H. Sung, J. Lim, J. Park, Y. Lee, Iris recognition using collarette boundary localization, in *Proceedings of International Conference on Pattern Recognition*, pp. 857–860 (2004)
16. J. Cui, Y. Wang, T. Tan, L. Ma, Z. Sun. A fast and robust iris localization method based on texture segmentation, in *Proceedings of the SPIE Defense and Security Symposium*, Vol. 5404, pp. 401–408 (2004)
17. W. Kong, D. Zhang, Accurate iris segmentation method based on novel reflection and eye-lash detection model. in Proceedings of International Symposium on Intelligent Multimedia, Video and Speech Processing, pp. 263–266 (2001)
18. P. Lili, X. Mei, The algorithm of iris image processing, in *Proceedings of 4th IEEE Workshop on Automatic Identification Technologies*, pp. 134–138 (2005)
19. C. Teo, H. Ewe, An efficient one-dimensional fractal analysis for iris recognition, *Proceedings of 13th WSCG International Conference in Central Europe on Computer Graphics, Visualization and Computer Vision*, pp. 157–160 (2005)
20. K. Grabowski, W. Sankowski, M. Zubert, M. Napieralska, Reliable iris localization method with application to iris recognition in near infrared light, *MIXDES (2006)*
21. X. He, P. Shi, A novel iris segmentation method for hand-held capture device, in *Springer LNCS 3832: International Conference on Biometrics*, pp. 479–485 (2006)
22. X. Feng, C. Fang, Z. Ding, Y. Wu, Iris localization with dual coarse-to-fine strategy, in *Proceedings of International Conference on Pattern Recognition*, pp. 553–556.(2006)
23. Q. Tian, Q. Pan, Y. Cheng, Q. Gao, Fast algorithm and application of hough transform in iris segmentation, in *Proceedings of International Conference on Machine Learning and Cybernetics*, vol. 7, pp. 3977–3980 (2004)
24. Y. Du, R. Ives, D. Etter, T. Welch, C. Chang, A new approach to iris pattern recognition. in *Proceedings of the SPIE European Symposium on Optics/Photonics in Defence and Security*, Vol. 5612, pp. 104–116 (2004)
25. T. Camus, R. Wildes, Reliable and fast eye finding in close-up images, in *Proceedings of International Conference on Pattern Recognition*, pp. 389–394 (2002)
26. D. Martin-Roche, C. Sanchez-Avila, R. Sanchez-Reillo, Iris recognition for biometric identification using dyadic wavelet transform zero-crossing. IEEE Aerosp. Electron. Syst. Mag. Mag. **17**(10), 3–6 (2002)
27. Chinese Academy of Sciences Institute of Automation, *Database of 756 Greyscale Eye Images*, http://www.sinobiometrics.com
28. P. Phillips, K. Bowyer, P. Flynn, Comments on the CASIA version 1.0 iris dataset, *IEEE Trans. Pattern Anal. Mach. Intell.***29**(10), 1869–1870 (2007)
29. H. Proenca, L. Alexandre. Iris segmentation methodology for non-cooperative recognition, in *Proceedings of IEE Conference on Vision, Image and Signal Processing*, Vol. 153, pp 199–205 (2006)
30. H. Proenc͜ and L. Alexandre, UBIRIS: Iris image database, (2004), http://iris.di.ubi.pt
31. B. Bonney, R. Ives, D. Etter, Y. Du, Iris pattern extraction using bit planes and standard deviations, in *Proceedings of 38th Asilomar Conference on Signals, Systems, and Computers*, Vol. 1, pp. 582–586 (2004)
32. X. Li, Modeling intra-class variation for non-ideal iris recognition, *In Springer LNCS 3832: International Conference on Biometrics*, pp 419–427 (2006)
33. A. Abhyankar, L. Hornak, S. Schuckers, Offangle iris recognition using bi-orthogonal wavelet network system, in *Proceedings of 4th IEEE Workshop on Automatic Identification Technologies*, pp. 239–244 (2005)
34. A. Abhyankar, S. Schuckers, Active shape models for effective iris segmentation, *In SPIE 6202: Biometric Technology for Human Identification III*, pp. H1–H10 (2006)
35. High contrast iris image database downloaded from: http://phoenix.inf.upol.cz/iris/download/
36. P. Yao, J. Li, X. Ye, Z. Zhuang, B. Li, Iris recognition algorithm using modified log-gabor filters, in *Proceedings of International Conference on Pattern Recognition*, pp. 461–464 (2006)

37. P. Zhang, D. Li, Q. Wang, A novel iris recognition method based on feature fusion, in *Proceedings of International Conference on Machine Learning and Cybernetics*, pp. 3661–3665 (2004)
38. Z. Sun, T. Tan, Y. Wang, Robust encoding of local ordinal measures: A general framework of iris recognition, in Proceedings of BioAW Workshop, pp. 270–282 (2004)
39. Z. Sun, Y. Wang, T. Tan, J. Cui, Cascading statistical and structural classifiers for iris recognition, in *Proceedings of International Conference on Image Processing*, pp 1261–1262 (2004)
40. C. Park, J. Lee, Extracting and combining multimodal directional iris features, *In Springer LNCS 3832: International Conference on Biometrics*, pp. 389–396 (2006)
41. L. Chenhong, L. Zhaoyang, Efficient iris recognition by computing discriminable textons, in *Proceedings of International Conference on Neural Networks and Brain*, Vol. 2, pp. 1164–1167 (2005)
42. C. Chou, S. Shih, W. Chen, V. Cheng, Iris recognition with multi-scale edge-type matching, in *Proceedings of International Conference on Pattern Recognition*, pp. 545–548 (2006)
43. W. Boles, B. Boashash, A human identification technique using images of the iris and wavelet transform. IEEE Trans. Signal Process. **46**(4), 1185–1188 (1998)
44. C. Sanchez-Avila, R. Sanchez-Reillo, Multiscale analysis for iris biometrics, in *Proceedings of IEEE International Carnahan Conference on Security Technology*, pp. 35–38 (2002)
45. E. Krichen, M. Mellakh, S. Garcia-Salicetti, B. Dorizzi, Iris identification using wavelet packets, in *Proceedings of International Conference on Pattern Recognition*, pp. 335–338 (2004)
46. J. Thornton, M. Savvides, B. Vijaya-Kumar, An evaluation of iris pattern representations, *In Biometrics: Theory, Applications, and Systems* (2007)
47. J. Kim, S. Cho, J. Choi, R. Marks, Iris recognition using wavelet features. J. VLSI Signal Process Syst. **38**(2), 147–156 (2004)
48. C. Tisse, L. Torres, M. Robert, Person Identification based on iris patterns, in *Proceedings of the 15th International Conference on Vision interface* (2002)
49. K. Miyazawa, K. Ito, T. Aoki, K. Kobayashi, H. Nakajima, An efficient iris recognition algorithm using phase-based image matching, in *Proceedings of International Conference on Image Processing*, pp. 49–52 (2005)
50. S. Hosseini, B. Araabi, H. Zadeh, Shape analysis of stroma for iris recognition, in *Springer LNCS 4642 International Conference on Biometrics*, pp. 790–799 (2007)
51. A. Azizi, H. Pourreza, A novel method using contourlet to extract features for iris recognition system, In Springer LNCS 5754, *International Conference on Emerging Intelligent Computing Technology and Applications*, pp. 544–554 (2009)
52. P. Hoyer, A. Hyvarinen, Independent component analysis applied to feature extraction from colour and stereo images. Netw. Comput. Neural Syst. **11**, 191–210 (2000)
53. P. Yuen, J. Lai, Face representation using independent component analysis, *Pattern Recogn.* **35**(6), 1247–1257 (2002)
54. B. Son, H. Won, G. Kee, Y. Lee, Discriminant iris feature and support vector machines for iris recognition, in *Proceedings of International Conference on Image Processing*, vol. 2, pp. 865–868, (2004)
55. M. Bartlett, J. Movellan, T. Sejnowski, Face recognition by independent component analysis, *IEEE Trans. Neural Netw.* **13**(6), 450–461 (2002)
56. H. Ekenel, B. Sankur, Feature selection in the independent component subspace for face recognition. *Pattern Recogn. Lett.* **25**(12), 1377–1388 (2004)
57. V. Dorairaj, N. Schmid, G. Fahmy, Performance evaluation of non-ideal Iris based recognition system implementing global ICA technique, in *Proceedings of ICIP*, vol. 3, pp. 11–14, (2004)
58. K. Bae, S. Noh, J. Kim, Iris feature extraction using independent component analysis, in *Proceedings. of 4th International Conference. Audio-and Video-Based Biometric Person Authentication*, Guildford, UK, vol. 2688, pp. 1059–1060 (2003)
59. K. Bowyer, K. Hollingsworth, P. Flynn, Image understanding for iris biometrics: a survey. *Comp. Vis. Image Understand., Acadamic Press,* **110**(2), 281–307 (2008)

60. I. Selesnick, R. Baraniuk, and N. Kingsbury, The dual tree complex wavelet transform: a coherent framework for multiscale signal and image processing, IEEE Signal Process. Mag. **22**(6), 123–151 (2005)
61. I. Selesnick, The design of approximate hilbert transform pairs of wavelet bases. IEEE Trans. Sig. Process. **50**(5), 1144–1152 (2002)
62. N. Kingsbury, The dual-tree complex wavelet transform: a new technique for shift invariance and directional filters, in *Proceedings of 8th IEEE DSP Workshop, Utah,* p. 86.20 (1998)
63. N. Kim, S. Udpa, Texture classification using rotated wavelet filters. IEEE Trans. Syst. Man Cybern. Part A Syst. Hum. **30**(6), 847–852 (2000)
64. M. Kokare, P. Biswas, B. Chatterji, Rotation invariant texture features using rotated complex wavelet for content based image retrieval, in *Proceedings* of *IEEE International. Conference on Image Processing, Singapore,* Vol. 1, pp. 393–396 (2003)

Chapter 3
Iris Segmentation

Abstract Two main challenges of iris segmentation of realistic eye images are processing speed and segmentation accuracy which have been addressed through two methods of iris segmentation in this chapter. The first method (using Canny edge detector) is primarily aiming at faster iris segmentation of more realistic images of UBIRIS database with sufficient segmentation accuracy. Second method is based on the pupil dynamics which has been used for fake iris detection but not for iris segmentation. This method has provided an accurate and loss-free iris segmentation. Unlike most of the other methods, this method does not assume circular or elliptical shape of iris. The performance of this method is tested on UPOL database images, which have high resolution and very low intensity gradient at iris boundaries. This novel method is also capable of fake iris detection.

Keywords Iris segmentation • Outer boundary localization • Pupil detection • Specular reflection removal • Iris normalization • Canny edge detection • Pupil dynamics • Customized iris segmentation • Accurate iris segmentation using pupil dynamics • CASIA database • UPOL database • UBIRIS database • Fake iris detection

3.1 Introduction

Human iris is unique and highly protected part of human being, and therefore, it has been researched extensively for personal authentication and identification. Human eye consists of not only iris but other parts such as pupil, sclera, eyelids, and eyelashes also. Iris is an annular part between sclera and pupil, and unique information required for the personal recognition is available only in this annular part, i.e., iris. Therefore, it is essential to extract the iris from eye image prior to feature extraction. The extraction of iris from eye image is called as iris segmentation or iris localization. It is a preprocessing stage of all iris

R. M. Bodade and S. N. Talbar, *Iris Analysis for Biometric Recognition Systems*,
SpringerBriefs in Computational Intelligence, DOI: 10.1007/978-81-322-1853-1_3,
© The Author(s) 2014

recognition systems. It is a very important step of any iris recognition system because not only it costs nearly half of recognition time [1] but also affects the recognition accuracy due to inaccurate iris segmentation. Therefore, improving speed of iris segmentation on one hand and accurate iris segmentation on other hand are the two important aspects of research on iris segmentation. Iris segmentation is not a simple issue in real-life conditions since iris is partially occluded by eyelids and eyelashes, specular reflections, pupil dilation, low contrast, unfocused images, low intensity gradients at outer and inner boundaries of iris, etc.

The related work on iris segmentation presented in Sect. 2.1 reveals that some of the issues of iris segmentation such as eyelid and eyelash occlusion, specular reflections, and low contrast and unfocussed images have been researched and addressed more or less. They are addressed either by proposing appropriate techniques to segment such images or by discarding the noisy images that are difficult to segment with the help of suitable image quality assessment schemes. Both the approaches add overheads in terms of increased processing time required for segmentation. As seen from the literature survey, most of the iris segmentation work is based on Wildes' [2–4] approach with some variations to address various issues. The literature survey on iris recognition shows that CASIA database [5] which is less realistic has been explored more than realistic databases such as UBIRIS [6] and UPOL [7] for the experimental results. As most of the existing methods of iris segmentation are tested on non-realistic, manually edited CASIA database, their performance cannot guarantee a similar amount of accuracy as for realistic images. Moreover, many of existing methods assume that iris and pupil are circular in nature and few assume it elliptical, which is seldom true, but they are actually of irregular shapes. Such methods always resulted in loss of unique and important *iris* data near the boundaries, especially in case of high-resolution images (e.g., images of UPOL database) having very low intensity gradient across both the boundaries of iris, the sclera–iris boundary (iris boundary or limbic boundary or sclera boundary or outer boundary) and the iris–pupil boundary (pupillary boundary or inner boundary).

Two main challenges of iris segmentation of realistic eye images are processing speed and segmentation accuracy which have been addressed through two methods of iris segmentation in this chapter.

The first method (using Canny edge detector) is primarily aiming at faster iris segmentation of more realistic images of UBIRIS database [6, 8] with sufficient segmentation accuracy to achieve the recognition rate as competent as achieved by the baseline segmentation methods, i.e., Wildes' method and Daugman's method.

Second method is based on the pupil dynamics which has been used for fake iris detection but not for iris segmentation. The primary aim of this method is an accurate and loss-free iris segmentation without prior assumption of fixed shape (circular or elliptical). The performance of this method is tested on UPOL [7] database images, which have high resolution and very low intensity gradient at iris boundaries.

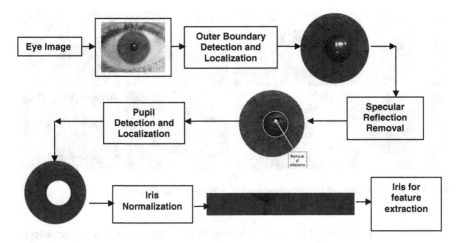

Fig. 3.1 Block diagram of proposed iris segmentation methods

3.2 Iris Segmentation

As shown in Fig. 3.1, the raw eye image contains not only iris but also the pupil, sclera, eyelids, eyelashes, specular reflections, etc., which are unwanted parts for iris recognition. The presence of these unwanted parts is considered as noise in iris recognition and adversely affects the performance of recognition. Therefore, an eye image has to be preprocessed to detect an iris. Following important steps are involved in both the proposed iris segmentation methods.

1. Outer boundary localization
2. Pupil detection
3. Specular reflection removal
4. Normalization.

We have devised two methods of iris segmentation, Method 1 is fast iris segmentation method, based on Canny edge detector which is customized for UBIRIS database and Method 2 is accurate iris segmentation method, based on pupil dynamics. Both methods are explained in the subsequent sections. Figure 3.1 shows the common block schematic for these methods.

3.3 Fast Iris Segmentation Method Using Canny Edge Detector Customized for UBIRIS Database

Daugman [9–12] makes use of an integro-differential operator for locating the circular iris and pupil regions, and the arcs of the upper and lower eyelids. However, the algorithm can fail where there is noise in the eye image, such as reflections, since it works only on a local scale.

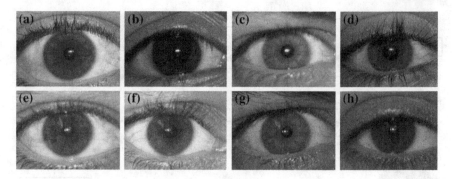

Fig. 3.2 Sample images of UBIRIS database [6]. **a** Session 1, **b** session 2

Wildes [2, 13] is based on contour fitting in two steps; firstly, an edge map is generated by calculating the first derivative of intensity values in an eye image using Laplacian operator, and then, binary edge map is obtained by thresholding the result. Secondly, from the edge map, votes are cast in Hough space, using circular Hough transform (CHT) [14] for the parameters of circles passing through each edge point using 'brute-force' approach.

Because of its better segmentation accuracy, this approach is used by many with some primary differences in edge detection operators; Yan et al. [15] use Robinson operator, Kang and Xu [16] use Sobel operator, and Ma et al. [17] use Canny operator. However, all these methods use Hough transform.

In spite of good segmentation accuracy, it has two major difficulties. It is very difficult to decide the value of threshold for iris segmentation (iris and pupil detection both) to produce the edge map of outer and inner boundary with the removal of edges of eyelid and eyelashes, especially in realistic eye images of UBIRIS database, shown in Fig. 3.2, in which variation of intensity across the whole image is not same. Therefore, two iterations are carried out; first to detect outer boundary and second to detect inner boundary. Moreover, the Hough transform is computationally intensive due to its 'brute-force' approach.

In this method, to overcome these problems of iris segmentation of realistic eye images, Canny edge detector [18] has been used for the detection of iris circle (outer boundary) only. Pupil detection is carried out separately by using efficient scanning methods. The parameters of iris circle (radius and center) are obtained using four tangents.

3.3.1 Empirical Study of UBIRIS Database

UBIRIS database [6] is composed of 1,877 images collected from 241 persons during September 2004 in two distinct sessions; Session-1 was carried out in a dark room for noise-free images; and Session-2 was carried out at different location in order to introduce natural luminosity factor. This enabled the appearance of heterogeneous images with respect to reflections, contrast, luminosity, and focus problems. It provides images with different types of noise, simulating image captured with or without minimal collaboration from the subjects, pretending to become an effective resource for the evaluation and development of robust iris identification methodologies.

The statistics of CASIA database is given below:

- Number of subjects: 241
- Number of images: 1,877 (1,205 in Session 1 and 672 in Session 2)
- Format: JPEG (color)
- Resolution: 800 × 600.

Figure 3.2 shows the sample images of UBIRIS database. From the empirical study of original images of UBIRIS database, following facts are noted and certain parameters are estimated for customization of segmentation algorithm.

- Images are captured in two sessions; Session 2 images are more noisy. Specular reflections are available in images of both sessions.
- Compared to CASIA image database, images of UBIRIS database are more realistic.
- In most of the images, iris is located at the center of an eye.
- Although iris is not exactly circular but slightly elongated in horizontal direction, it can be approximated as a circular ring between two circles, the iris circle (outer circle) and the pupil circle (inner circle).
- Range of iris radii varies from 100 to 130 pixels.
- Range of pupil radii varies from 40 to 65 pixels.
- Upper and lower limits for eye center are 140 and 100, respectively.
- Sclera, the outer part of iris, is bright for most regions, and pupil is the dark most region of an eye.
- Generally, only upper eyelid/eyelash occludes the upper portion of iris.

3.3.2 Outer Boundary Detection Using Canny Edge Detector

As stated above, to overcome the problems of Hough transform, Canny edge detector has been used for outer boundary detection because Canny edge detector is an optimal detector in which the edges are marked at maxima in gradient magnitude of a Gaussian-smoothened image.

3.3.2.1 The Canny Edge Detector

It is optimized edge detector, which detects true edges from noisy images and suppresses the false edges due to noise in an image. Therefore, this edge detector is more suitable for iris segmentation (outer boundary detection) of noisy eye images. Following four steps are involved in Canny edge detector [18].

1. *Smoothing using Gaussian filter* of standard deviation of 1.4 to minimize high-frequency noise that occurs mainly due to eyelid and eyelashes in eye images.
2. *Finding gradients* (magnitude and direction), using Sobel operator to mark edges where the gradients of the image has large magnitudes.
3. *Non-maximum suppression* to convert the 'blurred' edges in the image of the gradient magnitudes to get 'sharp' edges so that all local maxima in the gradient image are preserved.
4. *Double thresholding*, to remove the weak edges (false edges) which are not connected to strong edges. The edge pixels remaining after the non-maximum suppression step are marked with their strength pixel-by-pixel. Some of these are true edges in the image, but some are false edges as they are caused by noise. To discriminate among them, two thresholds are used. Edge pixels stronger than the high threshold are marked as strong edge pixel, and edge pixels weaker than the low threshold are removed but edge pixels between these two thresholds are marked as weak. Edge tracking is carried out to find which weak edges are connected to strong edges using 8-connectivity. Thus, all weak edges, which are not connected to strong edges, are removed and all other edges are preserved. This method is therefore less likely than others to be fooled by noise, and more likely to detect true weak edges.

3.3.2.2 Algorithm to Detect Outer Boundary

Following steps are involved in the detection of outer iris boundary using Canny edge detector.

1. Read raw colored eye image from UBIRIS database.
2. Convert this image to grayscale image to limit the computational requirements, as we are not exploring color information of iris for recognition.

$$I(x, y), \quad x = 0 \text{ to } 799, y = 0 \text{ to } 599 \tag{3.1}$$

3. Resizing the image to 256 × 256 for further reducing the computational overheads in processing to reduce the overall processing time.

$$I_1(x, y), \quad x = 0 \text{ to } 255, y = 0 \text{ to } 255 \tag{3.2}$$

4. Obtain threshold from histogram of an image. This step is optional just to check whether the automatically selected thresholds by Canny edge detector provide acceptable results or not.
5. In this method, $I_1(x, y)$ is the resized grayscale image of 256 × 256 obtained from original colored image of 800 × 600. Apply Canny edge detector with preset threshold value. It returns the binary image of same size as the input image, $I_1(x, y)$, with 1's where the function finds edges in $I_1(x, y)$ and 0's elsewhere. $I_{Canny}(x, y)$ is the binary edge map image. The sample grayscale images and their binary edge map images are shown in Fig. 3.3.
6. Obtain the two rows and two columns in the matrix of binary edge map image, $I_{Canny}(x, y)$, containing single '1' to draw the four tangents through these four boundary points. The details of this step are as given below.

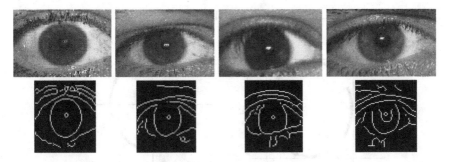

Fig. 3.3 Edge map of an eye image using Canny edge detector

Fig. 3.4 The part of matrix of edge map image

0	0	0	1	0	0
0	0	1	0	0	1
0	1	0	0	0	0
0	0	1	0	0	1
0	0	1	1	0	0
0	0	0	0	1	0

(a) The part of matrix of edge map image is shown in Fig. 3.4. For detecting the outer iris boundary, this matrix is read row-wise and column-wise within the range 65 and 195 to find a row and a column having a single '1' with rest of the pixels being '0', within the above-specified range. The rows and columns in the range of 65 and 195 are only analyzed instead of all rows and columns to reduce the time, as valid iris boundary lies within this range. Four such pixels (points), two each on two rows and two each on two columns, are determined.

(b) Then, four tangents are drawn through these points as shown in Fig. 3.5a.

7. By joining the points of contact of the two opposite tangents, two diameters are obtained, D_H and D_V. The diameter of iris is computed by taking the average of these two diameters.

$$D = \frac{(D_H + D_V)}{2} \tag{3.3}$$

Therefore, radius of outer boundary is $R = D/2$.

8. The point of intersection of these diameters is taken as the center of the iris as shown in Fig. 3.5b.

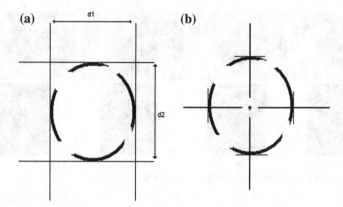

Fig. 3.5 Estimation of center and radius of outer boundary of an iris. **a** Four tangents, **b** iris center and outer radius detection

9. The iris center and the radius thus obtained are rough estimates of the outer circle of iris. Now, the region exterior to this circle is cropped by plotting 512 points on a circular arc with the help of obtained center and radius of iris. Figure 3.6 shows the accurately cropped iris images with pupil after removal of region exterior to the outer iris circle.

The cropped irises as shown in Fig. 3.6 are almost noise free, very less eyelashes noise is observed in very few images. Thus, Canny edge detector not only helps in detecting outer boundary accurately but also minimizes the noise. Thus, this method detects outer boundary more accurately with reduced processing time.

3.3.3 Pupil Detection and Localization

Once the iris has been separated from the rest of the eye, next step is to remove the pupil. Pupil is the darkest portion located at the middle of the eye. Therefore, the middle portion of the eye within the defined limits is scanned for pixels with intensity less than 60. This particular threshold is an approximation based on the analysis of the iris database, and its variation may give different results. The process of pupil detection and localization consists of three steps: removal of specular reflections, finding pupil center and radius, and pupil removal.

3.3.3.1 Specular Reflection Removal

Pupil is the dark most region of an eye located centrally, but the most bright spot is observed in this area, as shown in the images of Fig. 3.7a, which are specular reflections of camera light. These reflections cause error in pupil detection. From the empirical study of images, it is found that these bright spots have intensity higher

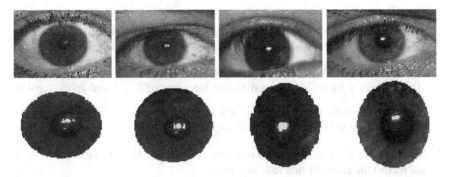

Fig. 3.6 Outputs of outer boundary detection and localization stage: extracted irises with pupil

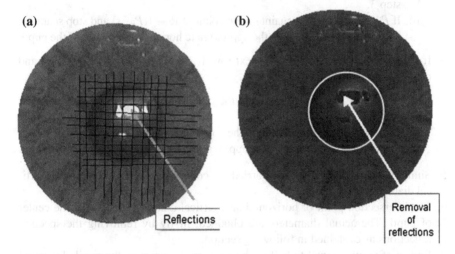

Fig. 3.7 Pupil detection and its removal to obtain iris ring. **a** Specular reflection, **b** removal of specular reflection and pupil detection

than 200. To avoid the error due to these reflections of light, the pixels of intensity above 200 in the area of interest (pupil area) are changed to a fixed value below 50 as shown in Fig. 3.7b and read as pupil area to include in the count during scanning of rows and columns. Thus, the correct center and radius of the pupil are obtained.

3.3.3.2 Finding Radius and Center of Pupil

In this method, we start scanning from central row and central column to record the number of pixels having intensity value less than 60 (which are the pixels of pupil region) to reduce the scanning time and in turn the pupil detection time. The steps of algorithm are as follows:

1. Let central row be R and central column be C and number of pixels having intensity less than 60 in this row and column be P_R and P_C, respectively.

2. Previous row, R − 1, and next rows, R + 1, are scanned, and number of such pixels recorded for them are P_{R-1}, P_{R+1}, respectively. These counts are compared with P_R, and highest among these three is recorded.

   (a) If $P_{R-1} > P_R > P_{R+1}$, then maintain the count of R − 1 and go to step 3.
   (b) If $P_{R+1} > P_R > P_{R-1}$, then maintain the count of R + 1 and go to step 4.

   Else, maintain the count of $R(P_R)$ and stop scanning rows. This count, P_R, is the approximate horizontal diameter of the pupil.

3. If $P_{R-1} > P_R > P_{R+1}$, then scan previous row, [(R − 1) − 1] i.e., scan R − 2 and record the count of this row, say, P_{R-2}.

   (a) If $P_{R-2} > P_{R-1}$, then maintain the count of R − 2, say P_{R-2} and repeat step 3.
   (b) If $P_{R-1} > P_{R-2}$, then maintain the count of R − 1(P_{R-1}) and stop scanning rows. This count, P_{R-1}, is the approximate horizontal diameter of the pupil.

4. If $P_{R+1} > P_R > P_{R-1}$, then scan next row, [(R + 1) + 1] i.e., scan R + 2 and record the count of this row, say, P_{R+2}.

   (a) If $P_{R+2} > P_{R+1}$, then maintain the count of R + 2, say P_{R+2} and repeat step 4.
   (b) If $P_{R+1} > P_{R+2}$, then maintain the count of R + 1(P_{R+1}) and stop scanning rows. This count, P_{R+1}, is the approximate horizontal diameter of the pupil.

5. Similarly, steps 2, 3, and 4 are carried for column scanning to obtain the vertical diameter of the pupil.
6. The intersection point of horizontal and vertical diameter of pupil is the center of pupil. The actual diameters are obtained, firstly by removing the specular reflections as explained in following section.
7. Radius of pupil is equal to half of the average diameters of the pupil. The pupil circle drawn with the help of parameters so computed is shown in Fig. 3.8a.

From the experimental results, it is found that average time required for pupil detection using this algorithm is less than half the time required for pupil detection using both the methods, i.e., conventional Hough transform and Daugman's method. Thus, this algorithm is time-efficient algorithm of pupil detection.

3.3.3.3 Removal (Cropping) of Pupil from Iris

Now, from the center of pupil, 512 points are plotted on a circle of pupil radius and the region interior to it is cropped from the iris with pupil image as shown in Fig. 3.8b.

It is observed that the pupil region is not concentric within the iris region and is usually slightly nasal, i.e., pupil center and *iris* center do not coincide with each other, but they are at different locations. This can be visualized from Fig. 3.9. This 'doughnut' shaped iris region is converted to constant size rectangular iris image by using the iris normalization process.

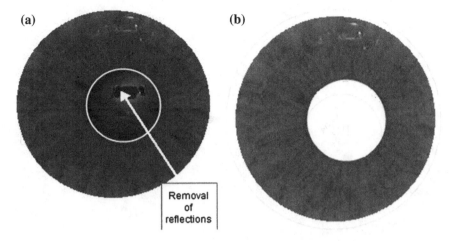

Fig. 3.8 Detection of pupil circle and removal of pupil

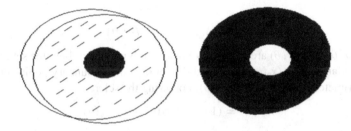

Fig. 3.9 Doughnut shaped iris region with iris center and pupil center at different locations

3.3.3.4 Iris Normalization

Iris normalization is a process of extracting a fixed number of pixels from circumference, with successive radius values of iris ring to form a normalized rectangular iris image of fixed dimensions to overcome the variations due to pupil dilation, scale variance, rotation variance, etc. which occurs due to imaging inconsistencies. Daugman described in [9–12], the rubber sheet model for iris normalization, which find the corresponding points in inner and outer iris boundaries and find some points in the line which is drawn between these two points to translate the captured data to a dimensionless polar coordinate system. Figure 3.10 illustrates the translation process, which has a polar (θ) and radial (r) variables.

In Fig. 3.10, 'i' and 'p' represent the center of the iris and center of the pupil, respectively, and (ox, oy) is the difference between both centers in x and y directions. The rubber sheet [9–12] is a linear model that assigns to each pixel of the iris, a pair of real coordinates (r, θ), where 'r' is on the unit interval [0, 1] and 'θ' is an angle in [0, 2π] to compensate the doughnut shape due to displaced centers using following equations.

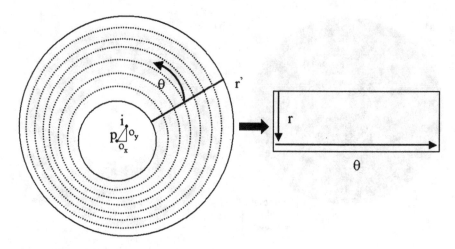

Fig. 3.10 Daugman's rubber sheet model of iris normalization

$$I(x(r,\theta), y(r,\theta)) \rightarrow I(r,\theta) \qquad (3.4)$$

where $x(r, \theta)$ and $y(r, \theta)$ are defined as linear combinations of both the set of pupillary boundary points $(x_p(\theta), y_p(\theta))$ and the set of limbic boundary points along the outer perimeter of the iris $(x_s(\theta), y_s(\theta))$ bordering the sclera:

$$x(r,\theta) = (1 - r) * x_p(\theta) + r * x_s(\theta) \qquad (3.5)$$

$$y(r,\theta) = (1 - r) * y_p(\theta) + r * y_s(\theta) \qquad (3.6)$$

Although these equations normalize the iris against all adversaries, these equations are complex for computations mainly due to different location of centers of pupil and iris as shown in Figs. 3.9 and 3.10.

In our method, we simplified these equations by considering pupil center at the position of iris center (call this as 'center of ring') and then crop the area to obtain the modified iris ring of uniform inner boundary of exactly circular shape as shown in Fig. 3.11. The ring so obtained may not be circular on outer boundary because the area exterior to iris already has been cropped and cropping of pupil is done with reference to center of ring and not with reference to center of pupil.

As both the centers coincide with each other, the equations of normalization are simplified but a small additional part of iris region is also removed along with the unwanted pupil. Literature survey [tan] revealed that by defining 'region of interest' of iris to remove the eyelid and eyelash noise from iris resulted in minimizing the recognition errors. Thus, the extra cropping of iris region to map pupil center to iris center is viewed as defining the 'region of interest' of iris not only for improving the processing speed, but also to reduce the noise.

To convert new modified iris ring of reduced dimensions obtained by coinciding the pupil center with iris center into polar plane of constant dimension,

Fig. 3.11 Removal of additional part of iris with iris center and pupil radius

Fig. 3.12 Iris ring and normalized iris image of size 512 × 32

normalization Eqs. (3.4)–(3.6) of Daugman model are simplified to following equations:

$$x_1 = x + r \cos \theta \qquad (3.7)$$

$$y_1 = y + r \sin \theta \qquad (3.8)$$

where (x, y) are the coordinates of center of the ring and (x_1, y_1) are the coordinates of points on the circle.

In this step, equal number of points have been chosen, clockwise in each boundary. By connecting each point on inner boundary to its corresponding point on outer boundary, iris pixel coordinates are mapped into a dimensionless polar space using the Eqs. (3.7) and (3.8). For this purpose, 32 values are taken between the inner circle radius and the outer circle radius and 512 points are plotted on each of these 32 circles. All these points are mapped row-wise to a rectangular matrix. Thus, the normalized iris of fixed size (512 × 32), shown in Fig. 3.12, is obtained by this simplified normalization process.

To further reduce the chances of error where ring is not complete, shown in Fig. 3.13, the matrix so obtained is analyzed for inconsistent values and that portion is removed from the image. For this purpose, the matrix is checked column-wise and a count of the columns which are consistent is maintained. Only the longest continuous pattern without any inconsistency is retained and repeated to complete the rectangle. The simple image quality assessment is done at this stage

Fig. 3.13 Example of incomplete iris ring having number columns more than 128

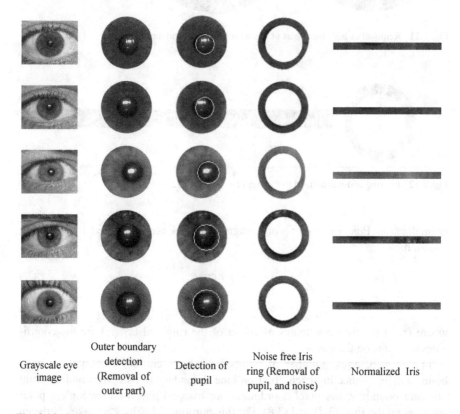

| Grayscale eye image | Outer boundary detection (Removal of outer part) | Detection of pupil | Noise free Iris ring (Removal of pupil, and noise) | Normalized Iris |

Fig. 3.14 Sample eye images of UBIRIS database and outputs at key stages of proposed iris segmentation method

by ensuring the presence of at least 128 columns for further processing of an iris image. If the columns are less than 128, then such image is labeled as bad image and it is disqualified for further processing. An incomplete iris ring having more than 128 columns and its normalized iris pattern of 512×32 is shown in Fig. 3.13. Finally, Fig. 3.14 shows the sample eye images and outputs at key stages of this method.

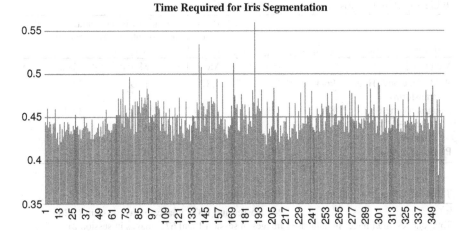

Fig. 3.15 Time required for iris segmentation, *red line* shows minimum time and *green line* indicates maximum time (Color figure online)

3.3.4 Experimental Results

Following two experiments are conducted to evaluate the performance of the iris segmentation algorithm in terms of segmentation accuracy, processing time required for segmentation, and recognition errors of recognition system when normalized iris image (with partial loss of iris data) obtained from this method are used as input for feature extraction and matching. The implementations are carried out in MATLAB7.0, on PIV-3Ghz, Intel processor with 512 MB RAM and tested on UBIRIS database to conduct following experiments.

3.3.4.1 Results of Iris Segmentation on UBIRIS Database

We have also implemented the iris segmentation methods using integro-differential operator (Daugman's model of iris segmentation) and Hough transform (Wildes' model of iris segmentation) by using the freely available open source MATLAB codes [19] with appropriate modifications for images of UBIRIS database, and we compared the segmentation accuracy and processing time achieved by our method with these methods. The actual time of iris segmentation of all eye images is measured, and it is plotted for 350 images as shown in Fig. 3.15. We obtained the binary edge map in case of Wildes' method using Canny edge detector. We used all 1,877 eye images of UBIRIS database for iris segmentation using these three methods, and the results of this experiment are tabulated in Table 3.1.

Table 3.1 Results of iris segmentation on UBIRIS database

Performance measure →	Segmentation accuracy	Processing time of iris segmentation (total time required to obtain normalized iris image from resized grayscale image)		
Methods ↓	(%)	Mean time (s)	Min. time (s)	Max. time (s)
Wildes' method (Hough transform)	96.4	1.87	1.69	1.97
Daugman's method (intergo-differential operator)	87.7	1.32	1.11	1.64
Proposed method 1	95.7	0.437	0.281	0.547

Table 3.2 Details of gallery set and probe set used for experiments

Total classes and images in database	241 classes and 1,877 images in two sessions (1,205 images in session 1 and 672 images in session 2)
Gallery size (for 100 classes)	500
Probe set size of genuine class	400
Probe set size of imposer class	777

3.3.4.2 Effect of Results of Iris Segmentation on Recognition Accuracy

We have also implemented the feature extraction and pattern-matching stages of Daugman's system by using freely available open source MATLAB codes [19] to evaluate the effect of partial loss of iris data in segmented iris of our method, on recognition rate. With the help of this implementation, iris codes [9–12] of gallery set images are generated and compared with iris codes of images of probe set to obtain recognition errors, using the segmented normalized irises obtained from our segmentation results and other two baseline methods as inputs to feature extraction and pattern-matching module.

We used 05 images per class (03 from Session 1 and 2 from Session 2 of UBIRIS database) for training and remaining images as test images; the details of gallery set and probe set of UBIRIS database used for this experiment is given in Table 3.2.

The ROC curves (FAR vs. TAR) are plotted, as shown in Fig. 3.16, to compare the performance of recognition. In this plot, blue colored curve is for segmented iris images of Daugman's approach, green colored curve is for segmented iris images of Wildes' approach, and red colored curve is for segmented iris images of our approach.

From ROC curve of Fig. 3.16, recognition accuracy (true accept rate) at FAR = 0.01 for the proposed method, Daugman's method, and Wildes' method are 89.4, 84.3, and 92.6 %, respectively.

3.3.4.3 Analysis of Experimental Results and Inferences

The analysis and inferences of results of above-stated experiments are stated in following paragraphs:

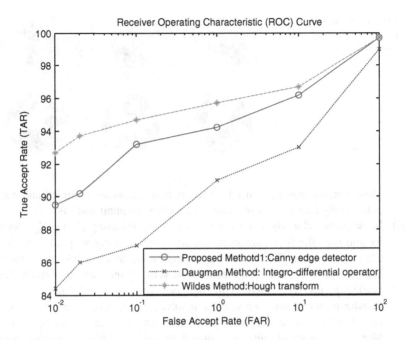

Fig. 3.16 ROC curves of our method, Daugman's method and Wildes' method

1. Time required for iris segmentation and normalization process is very less for proposed method (0.437 s) as compared to Daugman's method (1.32 s) and Wildes' method (1.87 s). The time efficiency achieved in the proposed method as compared to other two methods is due to the following reasons:

   (a) In Wildes' method, two steps are involved in outer boundary detection: The first step is to convert the gray-scaled eye image into a binary edge map using suitable (Canny) edge detection method, and then, CHT is used. In the CHT, each edge pixel with coordinates (x, y) in the image space is mapped for the parameters space determining the circle parameters (radius: r and coordinates of center: x_c, y_c) which resolves the circle equation. As a result, the point (x_c, y_c, r) is obtained in the parameters space which represents a possible iris circle in the image, all such possible circles are obtained, and then correct iris circle is drawn. As intensity levels, center and circumference of pupil is different from that of iris, hence, these two steps are repeated for the detection of pupil. This makes this method computationally more heavy, and hence, time required for iris segmentation using this method is highest among three. However, the accuracy of segmentation with minimal loss iris data is best, among all the three methods.

   (b) In Daugman's method, integro-differential operator is used for the detection of both the boundaries, one after other. This operator searches for the circular path where there is a greatest change in pixel values when there is a variation of the radius and the x and y coordinates from the center of the iris circle in iterative operations. The same process is repeated for pupil circle.

Fig. 3.17 Inaccurate iris
segmentation from noisy
images using Daugman's
method

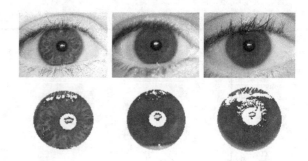

Moreover, the equations used for iris normalization are complex and computationally heavy because of different centers of pupil and iris.

(c) In the proposed method, outer boundary is detected using Canny edge detector and then the four tangent points are obtained using very fast search of edge map matrix. Finally, the center and radius of iris circle are obtained from the cross-section of horizontal and vertical lines drawn from the four tangent points. This method neither uses computationally heavy Hough transform nor iterative operation involved in it. Unlike existing methods, pupil is detected separately by using very efficient scanning method. Moreover, center of pupil is mapped to center of iris, to remove noise and some part of iris for simplifying the normalization equations. Finally, empirical study of UBIRIS database is made to customize this method for UBIRIS database. Thus, in this method, processing time is saved in all steps, outer boundary detection, pupil detection, and normalization. Therefore, this method is computationally very efficient and requires very less time for iris segmentation and normalization as compared to other two methods.

2. The normalized iris obtained using Wildes' method is complete and with minimal loss of useful iris data, whereas, in our method, some useful part of iris is also lost, and in few cases, iris ring is also incompetent due to mapping of pupil circle to iris circle to simplify the equations of normalization to remove the noise. This is also a reason of slightly less segmentation accuracy in our method as compared to Wildes' method. The effect of noise is negligible in the proposed method and Wildes' method because of use of Canny edge detector which removes the weak edges of noise. As Daugman's method works only on local scales, it is incapable to handle noise in realistic eye images and it requires additional mechanism to suppress the noise, especially, reflections in noisy images. Therefore, success rate of iris segmentation is poor in this method. Figure 3.17 shows the effect of eyelid and reflection noise in the process of iris segmentation using Daugman's method.

3. Recognition accuracy (TAR at FAR = 0.01 %) using iris images of Wildes' method is maximum (92.6 %), that in case of our proposed method is slightly below it (89.4 %) and it is minimum (84.3 %)in case of Daugman's method. Following inferences are drawn from these results.

(a) The best recognition performance using images of Wildes' method is not only due to the highest success rate of iris segmentation but more

(a)   **(b)**   **(c)**   **(d)**

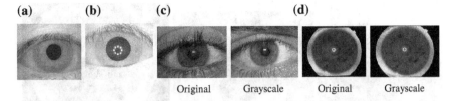

Original   Grayscale   Original   Grayscale

Fig. 3.18 Comparison of images of CASIA [5], UBIRIS [6], and UPOL [7] databases. **a** CASIA ver1, **b** CASIA ver3, **c** UBIRIS ver3, **d** UPOL

      importantly due to accurate iris segmentation of complete iris with minimal loss of iris data.

(b) In our method, although iris segmentation rate is comparable to Wildes' method but the recognition performance is inferior because of loss of iris part during pupil removal. To match the recognition accuracy of Wildes' method, feature extraction algorithm of better performance need to be introduced.

(c) This indicates that the *iris* data close to pupil boundary is very unique and useful and loss of this part of iris adversely affects the recognition performance.

(d) The least recognition rate in case of Daugman's method is obvious because of low success rate of iris segmentation and presence of noise in segmented iris images.

3.4 Accurate Iris Segmentation Method Using Pupil Dynamics

In the iris segmentation method discussed in Sect. 3.3, we are able to achieve the fairly good success rate of iris segmentation but the recognition accuracy using these iris images is much inferior, because of loss of iris data (iris part), especially, near pupil boundary. Any loss of iris part in the segmentation process, in turn, weakens the very useful unique iris features, which are essential for differentiating interclass subjects, thus adversely affects the recognition performance of system. This fact is also supported by the results of our proposed method explained in earlier section. Also, Ma et al. [20] reported that region close to pupil boundary contains most unique features of iris and noise interference (due to eyelid and eyelashes) to this part is also negligible literature survey on iris segmentation reveals that in most of the methods, iris and pupil are considered as circular in shape, which is seldom true. Therefore, loss of iris part, less or more, is obvious in such iris segmentation methods. To minimize this loss, few [21–24] assumed it elliptical rather than circular and reported improvement in the recognition results but at the higher computational cost as more parameters are required to define ellipse. Although ellipse is better shape descriptor of iris than circle, an iris is of irregular shape, in actual.

Figure 3.18 shows images of CASIA version 1, CASIA version 3, UBIRIS version 3, and UPOL database. Following points are observed from the study of these images:

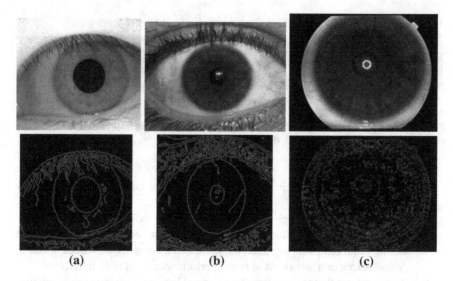

Fig. 3.19 Binary edge map of images of CASIA, UBIRIS, and UPOL databases. **a** CASIA, **b** UBIRIS, **c** UPOL

(a) CASIA images are of low resolution, and UBIRIS images and UPOL images are of high resolution.

(b) Gradient of intensity across outer boundary and across inner boundary is very high in CASIA images, moderate in UBIRIS images, and low in UPOL images, i.e., outer boundary and inner boundary are least visible in UPOL image and very clearly visible in CASIA image.

(c) Reflections of flashlight is observed, more prominently in images of CASIA and UPOL database as compared to UBIRIS image, and there is no reflection in image of CASIA version 1. Moreover, the reflections in images of CASIA version 3 databases are controlled well within the substantially dark and clearly defined pupil region, whereas reflections in the images of UPOL database look more realistic and challenging to address.

(d) Effect of eyelid and eyelashes is more in CASIA as compared to other database, but various successful techniques [25–28] have been proposed to overcome this problem.

(e) Iris and pupil are not exactly circular in shape.

(f) The binary edge map using Canny edge detector of images of CASIA, UBIRIS, and UPOL database, as shown in Fig. 3.19, edge map of UPOL database image do not reveal significant boundary which can support shape-based iris segmentation using conventional segmentation methods.

Thus, the result of our proposed Method 1, literature survey and study of eye image databases have motivated us to devise new iris segmentation method for accurate segmentation. This method shall be capable of segmenting iris of actual shape, without any prior assumption of fixed shape or any loss of unique iris data (especially near inner boundary) from high-resolution images, which are having low intensity gradient at the boundaries (images of UPOL database) at comparable processing time. This method is based on pupil dynamics. Variation in pupil size due to change in illumination is termed as pupil dynamics. This technique has been used for fake *iris* detection [28–32] to safeguard the iris recognition system from attack of forged irises. However, to the best of our knowledge, pupil dynamics have yet not been used for iris segmentation. Therefore, this method is also capable of detection of certain types of fake irises [32].

3.4.1 Flowchart of Proposed Method

Size of pupil automatically changes if the light intensity incident to eye is changed to control the amount of light entering inside the eye. This is called as pupil dilation. Thus, size of outer boundary is fixed and remains constant, but inner boundary (pupil size) varies with illumination. This principle of pupil referred hereafter as, '*pupil dynamics*,' is explored for iris segmentation method. The proposed method uses two images of same subject acquired at different intensities of light to detect the changes in size of pupil (pupil dynamics).

The complete overview of the proposed system is represented by a flowchart as shown in Fig. 3.20. The system mainly consists of preprocessing, outer boundary detection, inner boundary detection, and normalization.

3.4.2 Preprocessing

Firstly, an original colored eye image of size 768×576 is converted to grayscale image. For selection of outer and inner boundaries, grayscale image is converted into binary image using Otsu's threshold [30]. Figure 3.21a–d shows an original eye image, its grayscale, binary, and inverted binary images, respectively.

3.4.3 Outer Boundary Detection

Existing algorithms assume iris images exactly circular in nature, which is seldom true, but such assumption results into failure of iris segmentation, in certain high-resolution images where intensity gradients across sclera–iris and iris–pupil is low. When high-resolution images, where intensity gradients across sclera–iris and iris–pupil are low, are converted into binary images, the resultant images are not in the form of circular objects, but they are actually in the form of irregular shapes. Hence, in this method, we are not tracing or searching for circular objects, but we have assumed that resultant images can be of any shape and structure.

Fig. 3.20 Flow graph of
proposed method 2

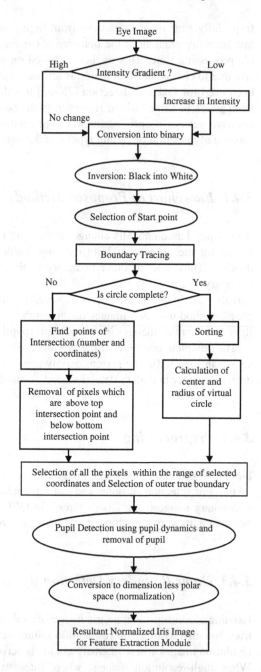

In this algorithm, binary image is traced and pixels are classified based upon
values of their intensities, i.e., one group with intensity of 1 (white) and other with
intensity of 0 (black) as shown in Fig. 3.21c. Then, binary image is inverted as
shown in Fig. 3.21d.

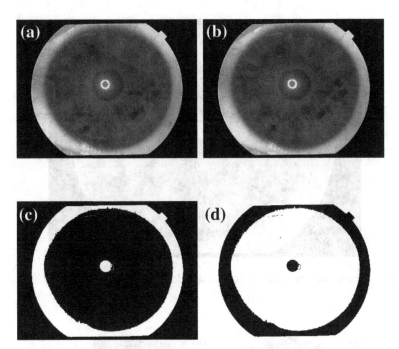

Fig. 3.21 Preprocessing of an eye image for iris segmentation. **a** Original eye image, **b** grayscale image, **c** binary image, **d** inverted binary image

Using 8-connectivity, boundary of binary image is traced for all white pixels in all directions where *starting seed-point pixel* is properly chosen. To reduce the search operation, binary image is divided into four quarters. The starting seed pixel is first 'black' pixel encountered while tracing from top to bottom in any one quarter of image. Thus, complete boundary is traced to reach back to starting point that completes one circle (closed path) in case of images having complete iris, i.e., without interference of eyelids. In this case, there is no intersection. Traced points are then rearranged, sorted, and mapped on grayscale to detect the actual outer boundary which is shown by green colored trace in Fig. 3.22. Approximate circle is also estimated from this actual trace by using back slash operator that gives coordinates of the center and radius. The estimated virtual circle (shown in blue color), and accurate actual outer boundary of Iris traced by using proposed method is shown by green color in Fig. 3.22. Center of virtual circle is shown by yellow dot at the center. These parameters are mapped on both the grayscale images (two images at two intensities) to localize the iris. Accurately localized irises of actual shape from both images are shown in Fig. 3.23. Both images have same outer boundary, but pupil of different size.

Figure 3.24 shows incomplete iris images due to intersection with upper or lower eyelids. For such images, tracing of outer boundary may not result into a complete one object (closed circular path); for such cases, point of intersection is calculated and all

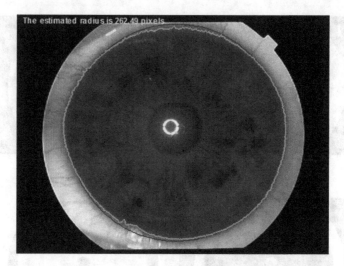

Fig. 3.22 Accurate outer boundary (of actual shape) detection of iris

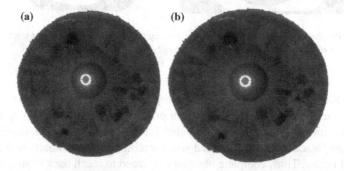

Fig. 3.23 Accurate outer boundary detection and localization from two images of different pupil size. Figure shows exactly similar outer boundary. **a** Image of smaller pupil, **b** image of larger pupil

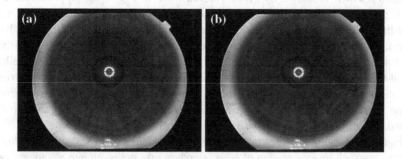

Fig. 3.24 Original image and grayscale image with intersection with *upper* eyelid

pixels above point of intersection. In case of intersection with upper eyelid and all pixels below the point of intersections in case of intersection with lower eyelids are removed.

In case of complete iris, area under the traced circle (green color) boundary is selected, and in case of intersection, area within virtual circle (blue color) boundary is selected. This selected area is cropped from rest of the image and copied to new image which is used for pupil detection stage. Similarly, second image of the same subject is also processed. Two different images of same subject acquired at two different illuminations, one after another, will have differences in the size of pupils with same outer boundaries. These images are used for pupil detection and fake iris detection using pupil dynamics.

3.4.4 Pupil Detection

Once the iris has been separated from the rest of the eye, next step is to remove the pupil. Pupil is the darkest portion near the center of the eye. So the middle portion of the eye within the limits defined is scanned for pixels with intensity less than 60. This particular threshold is an approximation based on the analysis of the iris database, and its variation may give different and incorrect results. Therefore, to avoid such variation, "two-image reference technique" is used for pupil detection in our method.

In this method, it is assumed that two images of same subject are acquired in a small interval of time (one after another) under different illuminations. The images of irises obtained from previous stage are converted to binary images. Then, second binary image is subtracted from the first one to detect the variation in size of pupil. As iris part of two images is same, result of subtraction will give zero value and only place where non-zero values are obtained, is the region of pupil due to variation in size of pupil.

3.4.5 Removal of Specular Reflections

The effect of specular reflection in pupils creates small bunch of black pixels inside the pupil as shown in Fig. 3.25a. These small regions of black pixels can be wrongly interpreted as pupil, and it will lead to incorrect pupil detection. Therefore, these small parts (regions) of pupil need to be removed. These small parts are much smaller than one-sixth of correct pupil. Therefore, exploiting the empirical information of database, any region having less than 30 black pixels is considered as reflection noise and removed from the image (converted to white pixels). This results into removal of reflection noise, and complete pupil is detected as shown in Fig. 3.25b.

Tracing this inner boundary and selecting region outside inner boundary and below outer boundary will give accurate iris of actual shape without loss of any iris data as shown in Fig. 3.26.

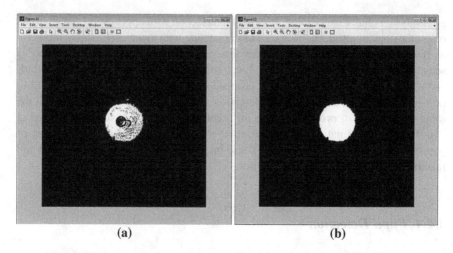

(a)   (b)

Fig. 3.25 **a** Pupil region with reflection noise and **b** removal of reflection noise

Fig. 3.26 Accurately
segmented iris of actual
shape without loss of any
iris data

3.4.6 Normalization

Finally, completely detected iris is converted to dimensionless fixed size rectangular image of fixed size using normalization Eqs. (3.4)–(3.6) of Daugman's rubber sheet model as presented in earlier Sect. 3.3.4.

In this step, the ring is converted to a more convenient rectangular image without losing the texture information. For this purpose, 128 values are taken between the inner circle radius and the outer circle radius and 512 points are plotted on each of these 128 circles. All these points are mapped row-wise to a rectangular matrix. Thus, a rectangle of size 512×128 is obtained.

Table 3.3 Details of pupil size variation to modify the original images

Pupil size variation (%)	Less than 5 % of original size	5 % of original size	10 % of original size	15 % of original size
Number of images (%)	30 % images	25 % images	25 % images	20 % images

3.4.7 Fake Iris Detection

As pupil detection is carried out by comparing the size of pupils of two images of same person, the difference in size of two pupils is measured. If variation is in the range of 5–15 %, then it is considered as real eye, else fake eye. Thus, this iris segmentation method is capable of detecting fake irises too.

3.4.8 Experimental Results

This method is implemented in MATLAB7.0, on PIV-3Ghz, Intel processor with 512 MB RAM and tested on all 384 images of UPOL database.

3.4.8.1 UPOL Database

The UPOL database contains 24 bit—RGB, $3 \times 128 = 384$ iris images (i.e., 3×64 left and 3×64 right) of 576×768 resolution in PNG format. These images are captured using TOPCON TRC50IA optical device connected with SONYDXC-950P3 CCD camera [31]. To simulate the variation in pupil size due to change in illumination, we used image-editing software and modified the images to change the size of pupil. Each modified image is paired with its original image and is used for iris segmentation of original image. The statistics of pupil variation obtained after modification is tabulated in Table 3.3.

3.4.8.2 Performance of Iris Segmentation Using Pupil Dynamics Method

In this experiment, we have evaluated the performance of the iris segmentation algorithm in terms of segmentation accuracy and processing time required for segmentation. Figure 3.27 shows the sample eye images and outputs at key stages of this method. This algorithm with appropriate modifications was also tested on few occlusion-free images of UBIRIS database. The result of this experiment is presented in Table 3.4. From Fig. 3.28, it is clear that Daugman's method and Wildes' method failed in iris segmentation from images of UPOL database.

Fig. 3.27 Normalized iris image without loss of any iris data

Table 3.4 Performance evaluation of iris segmentation using pupil dynamics method

Performance measure →	Segmentation accuracy (%)	Processing time of iris segmentation (total time required to obtain normalized iris image)		
Database ↓		Mean time (s)	Min. time (s)	Max. time (s)
UPOL	99.47	1.172	1.135	1.207
UBIRIS (occlusion-free images)	100	1.688	1.645	1.753
Average of (UPOL and UBIRIS)	99.73	1.43	1.39	1.48

3.4.8.3 Effect of Accurate Iris Segmentation on Recognition Accuracy

We have also tested the recognition performance of Daugman's system as implemented in previous section by using accurately segmented iris images of UPOL database and plotted the ROC curve as shown in Fig. 3.28. ROC curves of Method 1, Daugman's method, and Wildes' methods are also shown in Fig. 3.28 for comparison with this method.

With the help of this implementation, iris codes [9–12] of gallery of set images are generated and compared with iris codes of images of probe set to obtain recognition errors, using segmented normalized irises obtained from above-stated segmentation method as inputs to feature extraction and pattern-matching module. We used 02 images per class for training and remaining images as test images, and the details of gallery set and probe set of UPOL database used for this experiment are given in Table 3.5.

From ROC curve of Fig. 3.29, recognition accuracy (true accept rate at FAR = 0.01) for the Method 1, Daugman's method, and Wildes' method are 89.4, 84.3, and 92.6 %, respectively, whereas for this novel method using pupil dynamics it is 94.8 %.

3.4.9 Analysis of Experimental Results of Pupil Dynamics Method

From the experimental results of above-stated experiments of pupil dynamics method based iris segmentation method, following inferences are drawn:

1. In this new method, time required for iris segmentation and normalization process on UPOL images and occlusion-free UBIRIS images is 1.172 and 1.688 s,

(a)   (b)   (c)   (d)

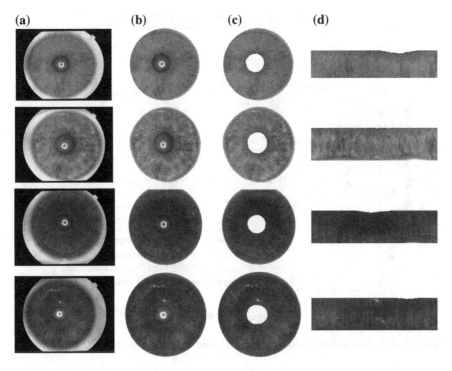

Fig. 3.28 Iris segmentation output for a few eye image. **a** Original eye image, **b** outer boundary detection (removal of outer part), **c** detection and removal of pupil, **d** normalized iris

Table 3.5 Gallery size and probe set size of UPOL database

Total classes and images in database	384 images of 128 classes (we considered left eye and right eye as separate class)
Gallery size (for 75 classes)	150
Probe set size of genuine class	75
Probe set size of imposer class	159

respectively. The average time required for images of both the databases together is 1.43 s. As compared to our first method, the time required in this new method is more but it is comparable to Daugman's method and it is less than Wildes' method, due to following reasons.

(a) This method is not based on any prior assumption of fixed shape but it traces the actual outer boundary of iris for accurate detection and localization of outer iris boundary. Moreover, to deal with the cases of intersection due to eyelids, additional processing is required. This accurate tracing of outer boundary requires more time as compared to the estimation of circle parameters in our first method and two baseline methods.

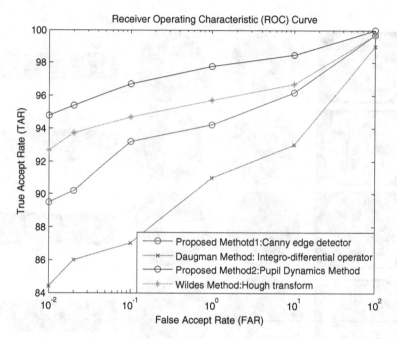

Fig. 3.29 ROC curves of our methods, Daugman's method and Wildes' method

(b) Secondly, for pupil detection and outer boundary detection of two images is carried out. This requires some additional time.

(c) In this method, as actual shape of outer and inner boundary is detected to segment an iris ring of actual shape and size, centroids of pupil and iris do not coincide, and hence, normalization process is based on Daugman's rubber sheet model. This also requires more time as compared to our first proposed method.

2. Pupil dynamics (variation in size of pupil with change in illumination) is used to detect the pupil and 5–15 % variation in pupil size with change in illumination is very important characteristics of live human eye. Thus, while detection of pupil, this method also checks whether difference in the pupil size of two images is less than 5 % of the pupil size for fake iris detection. This adds negligible overhead in terms of processing time but make iris recognition system robust against attacks of fake irises.

3. It is experimented that low intensity gradients at the boundaries (weak boundaries) of high-resolution images make iris segmentation extremely challenging and conventional iris segmentation methods failed in segmenting iris from such images.

4. Recognition accuracy (94.8 % at FAR = 0.01), using the normalized iris images of this new method, is highest among all the methods discussed in this research due to the following reasons.

(a) The success rate of iris segmentation using this method is close to 100 %.
(b) The best recognition performance using images of this method is not only due to the excellent success rate of iris segmentation but more importantly due to very accurate iris segmentation of complete iris of actual shape with negligible (almost lossless) loss of iris data (highly unique iris data near pupil region).

3.5 Summary

From the results of experiments and analysis of it thereon, following conclusions are drawn.

1. Only success of iris segmentation is not important, but the accuracy of segmented iris in terms of loss of *iris* data and interference of noise is also important for improving the recognition rate.
2. Best recognition rate is achieved with the help of our second method because of accurate segmentation of lossless and noise-free irises of actual shape with near 100 % success rate of segmentation.
3. Nowadays, choice of various high-resolution imaging devices and high-resolution images are available for recognition system. However, reduction in its resolution for ease of processing is generally done but this resizing of images to reduce the resolution adversely affects the recognition rate to some extent.
4. Processing time required for our first method is very less as compared to other methods discussed in this study. However, presented timings cannot be trusted completely because the processing time of algorithm depends not only on the power of the algorithm but on various factors such as way of implementation (loops and iteration used, code optimization, etc.), complexity in function of MATLAB, and overheads of other running processes on PC during execution of algorithm, and these aspects cannot be normalized in practice. Therefore, the processing time measured and stated in this study is only an indicator to compare the processing speed of algorithms and not the actual time required for the optimized computations of any algorithm.

However, the strength of algorithms in terms of processing time can very well be estimated from the fact that algorithms like Hough transform and Rubber sheet model of normalization require much more computations than implementation of simple basic logics like efficient raster scan method and use of tangent methods employed in our algorithm.

5. Although efficiency in processing time is essential for real-time recognition systems, it cannot be the prime aim of any recognition system, especially, in authentication systems where achieving "zero FAR" is the issue of highest concern. Moreover, reduction in processing time can be carried out by efficient implementations on specially designed processors as well. Accurate iris segmentation of iris with better (near 100 %) success rate of segmentation is more important than achieving a processing efficiency for obtaining the higher recognition rate.

6. The recognition rate using the very good quality noise-free and loss-free iris images obtained from *iris* segmentation method having near 100 % success rate is limited to 94.8 %. Extracting features using Gabor filter and using Hamming distance for matching have provided these results. This result reveals ample amount of scope of research in improving methods of feature extraction and matching to improve the recognition rate. This has motivated us to device new techniques of feature extraction which is as good as Daugman's method (Gabor filter based) but computationally efficient as presented in further chapters of this book.

References

1. X. Ye, Z. Zhuang, Y. Zhang, A new fast algorithm of iris location. Comput. Eng. Appl. **30**, 54–56 (2003)
2. R. Wildes, Iris recognition: an emerging biometric technology. Proc. IEEE **85**(9), 1348–1363 (1997)
3. R. Wildes, J. Asmuth, S. Hsu, R. Kolczynski, J. Matey, S. Mcbride, Automated noninvasive iris recognition system and method. U.S. Patent no. 5572596, 1996
4. R. Wildes, J. Asmuth, G. Green, S. Hsu, R. Kolczynski, J. Matey, S. McBride, A machine-vision system for iris recognition. Mach. Vis. Appl. **9**, 1–8 (1996)
5. Chinese Academy of Sciences Institute of Automation: Database of 756 greyscale eye images, http://www.sinobiometrics.com
6. H. Proenc, L. Alexandre, UBIRIS: Iris image database (2004), http://iris.di.ubi.pt
7. High contrast iris image database downloaded from: http://phoenix.inf.upol.cz/iris/download/
8. H. Proenca, L. Alexandre, Iris segmentation methodology for non-cooperative recognition, in *Proceedings of IEEE Conference on Vision, Image and Signal Processing*, vol. 153 (2006), pp. 199–205
9. J. Daugman, High confidence visual recognition of persons by a test of statistical independence. IEEE Trans. Pattern Anal. Mach. Intell. **15**(11), 1148–1161 (1991)
10. J. Daugman, Biometric personal identification system based on iris analysis. U.S. Patent no. 5,291,560, 1994
11. J. Daugman, The importance of being random: statistical principles of iris recognition. Pattern Recogn. **36**, 279–291 (2003)
12. J. Daugman, How iris recognition works. IEEE Trans. Circuits Syst. Video Technol. **14**(1), 21–30 (2004)
13. E. Wolff, *Anatomy of the Eye and Orbit*, 7th edn. (H. K. Lewis & Co LTD, London, 1976)
14. J. Illingworth, J. Kittler, A survey of the Hough transform. Comput. Vis. Graph. Image Process. **44**, 87–116 (1998)
15. M. Yan, X. Qi, Y. Wang, Research on a neural algorithm of iris recognition. Acta Biophys. Sin. **16**(4), 161–171 (2000)
16. H. Kang, G. Xu, Iris recognition system. J. Circuit Syst. **15**, 11–15 (2000)
17. L. Ma, Y. Wang, T. Tan, Iris recognition using circular symmetric filters, *in Proceedings of the 25th International Conference on Pattern Recognition (ICPR02)*, vol. 2 (2002), pp. 414–417
18. J. Canny, A computational approach to edge detection. IEEE Trans. Pattern Anal. Mach. Intell. **8**(6), 679–698 (1986)
19. L. Masek, Recognition of human iris patterns for biometric identification. Master's thesis, University of Western Australia, 2003. Available on: http://www.csse.uwa.edu.au/~pk/student projects/libor/LiborMasekThesis.pdf

20. L. Ma, T. Tan, Y. Wang, D. Zhang, Personal identification based on iris texture analysis. IEEE Trans. Pattern Anal. Mach. Intell. **25**(12), 1519–1533 (2003)
21. B. Bonney, R. Ives, D. Etter, Y. Du, Iris pattern extraction using bit planes and standard deviations, *in Proceedings of 38th Asilomar Conference on Signals, Systems, and Computers*, vol. 1 (2004), pp. 582–586
22. X. Li, Modeling intra-class variation for non-ideal iris recognition, *in Springer LNCS 3832: International Conference on Biometrics* (2006), pp. 419–427
23. A. Abhyankar, L. Hornak, S. Schuckers, Offangle iris recognition using bi-orthogonal wavelet network system, *in Proceedings of 4th IEEE Workshop on Automatic Identification Technologies* (2005), pp. 239–244
24. A. Abhyankar, S. Schuckers, Active shape models for effective iris segmentation, *in SPIE 6202: Biometric Technology for Human Identification III* (2006), pp. H1–H10
25. Y. Liu, S. Yuan, X. Zhu, Q. Cui, A practical iris acquisition system and a fast edges locating algorithm in iris recognition, *in Proceedings of IEEE Conference on Instrumentation and Measurement Technology* (2003), pp. 166–168
26. H. Sung, J. Lim, J. Park, Y. Lee, Iris recognition using collarette boundary localization, *in Proceedings of International Conference on Pattern Recognition* (2004), pp. 857–860
27. J. Cui, Y. Wang, T. Tan, L. Ma, Z. Sun, A fast and robust iris localization method based on texture segmentation, *in Proceedings of the SPIE Defense and Security Symposium*, vol. 5404 (2004), pp. 401–408
28. W. Kong, D. Zhang, Accurate iris segmentation method based on novel reflection and eyelash detection model, *in Proceedings International Symposium on Intelligent Multimedia, Video and Speech Processing* (2001), pp. 263–266
29. M. Clynes, M. Cohn, Color dynamics of the pupil. Ann. N.Y. Acad. Sci. **156**, 931–950 (1969), John Wiley Online Library, 2006
30. N. Otsu, A threshold selection method from gray-level histograms. IEEE Trans. Syst. Man Cybern. **9**(1), 62–66 (1979)
31. J. Daugman, Iris recognition and anti-spoofing countermeasures, in *Proceedings of 7th International Conference on Biometrics, London* (2004)
32. A. Pacut, A. Czajka, Aliveness detection for iris biometrics. IEEE International Carnahan Conference on Security Technology, 40th Annual Conference, October 17–19, Lexington, Kentucky, IEEE 2006, pp. 122–129 (2006)

Chapter 4
Iris Recognition Using Dual-Tree Complex Wavelet Transform and Rotated Complex Wavelet Filters

Abstract Iris is a rich-textured image having randomly oriented singularities (edges) in all directions and frequencies. Hence, Gabor wavelet-based represen-tation of iris images is an excellent solution. However, non-orthogonal band-pass filters tuned to one particular frequency and one particular orientation based on representation are computationally inefficient. On the other hand, discrete wavelet transform (DWT) is orthogonal and computationally efficient but provides edges only in three orientations which are not sufficient to preserve a uniqueness of an iris. Therefore, to overcome this limitation of DWT and exploits its strength of com-putational efficiency, the use of dual-tree complex wavelet transform (DT-CWT) and rotated complex wavelet filters (RCWF) has been used for feature extraction. The detail analysis of Gabor filters and DWT has been presented in this chapter. Theoretical aspects of wavelet transform and design of DWT filters, DT-CWT and RCWF, has been described in this chapter. Implementation issues and analysis of iris with adequate experimentations and deliberate comparison with existing algo-rithms are the key points of this chapter.

Keywords Feature extraction • DWT • Dual-tree complex wavelet transform • DT-DWT • Complex wavelet transform • CWT • Rotated complex wavelets • RCWF • Gabor wavelets • Limitations of DWT • Analytic wavelets • Multi-resolution analysis • Shift invariance test of DT-CWT • Canberra distance • Inter-class–intra-class separation test • Design of complex wavelet and rotated complex wavelet filters

4.1 Introduction

As brought out in Chap. 2, Daugman used Gabor filters for feature extraction and represented it with the help of binary iris code [1–4], whereas Wildes [5] made use of Laplacian of Gaussian filters at multiple scales to represent iris with a real-valued feature vector. Either of the approaches, binary representation of iris or real-valued feature vector of iris, has been explored very extensively by many researchers, mainly either by using variants of Gabor filters or by using discrete wavelet transform (DWT) for multi-resolution representation of iris.

Iris is a rich-textured image having randomly oriented singularities (edges) in all directions and frequencies. Gabor filter is a non-orthogonal band-pass filters tuned to one particular frequency and one particular orientation. Thus, multiple Gabor filters are required for extraction of iris features in multi-orientations and multi-resolutions. A central issue in applying these filters is the determination of the filter parameters. If the filter parameters are preset, then they are not necessarily optimal for a particular processing task [6, 7].

Apart from many techniques proposed in the literature, Gabor wavelet-based representation of iris images is an excellent solution. However, Gabor wavelet-based representation is computationally very complex and the memory requirement for storing Gabor features is very high. These problems were tried to overcome using sub-Gabor [8], simplified Gabor wavelets [9], optimal sampling of Gabor features [10], etc. But none of these approaches succeeded effectively. An approach to both shift invariance and directional selectivity was given by Simoncelli et al. [11].

On the other hand, DWT is orthogonal and computationally efficient because DWT is implemented using sub-band filters of real coefficient filters associated with real wavelets resulting in one real-valued approximation and three different sub-band details; i.e., it resolves an image into only three spatial-domain feature orientations: horizontal, vertical, and diagonal. This poor directionality affects the optimal representation of iris image. Moreover, DWT, being real-valued transform, cannot provide local phase information. Therefore, it is shift sensitive and weak in providing the exact locations of singularities.

To avail the local phase information, complex-valued filtering is required [12, 13] for the formulation of complex-valued 'analytic filters' or 'analytic wavelets' [14] that helps to get the required phase information, more orientations, and reduced shift sensitivity. Various approaches of filter bank implementation employing analytic filters associated with analytic wavelets are commonly referred to as 'complex wavelet transforms' (CWT). Selesnick et al. [15], Selesnick [16], and Kingsbury [17] introduced efficient method of CWT implementation using two trees of real-valued wavelet filters called dual-tree DWT or DT-CWT. In addition, Kokare et al. [18] introduced the rotated complex wavelet filters (RCWF) for obtaining the features in additional orientations from texture images by extending the basic concept of DWT rotation described by Kim and Udpa [19]. The formulations, design, and implementation of CWT filters and RCWF for iris feature extraction are discussed in this chapter.

4.2 Theoretical Aspects of Wavelet Transform

Fourier transform (FT) fails to correctly represent non-stationary signals. This limitation of FT is overcome by short-time Fourier transform (STFT), but due to preset size of window, it can achieve either time resolution or frequency resolution

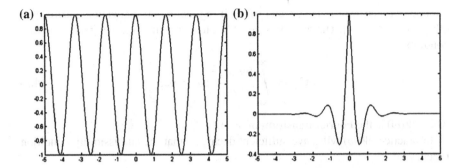

Fig. 4.1 Composition of wavelet using **a** a wave and **b** a wavelet

but not both. The arbitrary time and frequency resolutions can be obtained with multi-resolution approach of wavelet transform (WT). It gives good time resolution but poor frequency resolution at high frequency and vice versa. It uses wavelets of finite energy as window functions of two parameters: scaling (inverse of frequency) and translation (shifting). Wavelet transform is represented in continuous and in discrete domain.

4.2.1 Wavelets

A *wave* is usually defined as an oscillating function of time or space, e.g., sinusoid, whereas a wavelet is a 'small wave,' which has its energy concentrated in time as shown in Fig. 4.1.

The term wavelet was originally used in the field of seismology to describe the disturbances that emanate and proceed outward from a sharp seismic impulse. Fourier analysis is a wave analysis which expands signals or functions in terms of sinusoids (or, equivalently, complex exponentials). Fourier analysis has proven to be extremely valuable, especially for periodic, time-invariant, or stationary phenomena. A wavelet representation, however, is versatile in analysis of transient, time-variant, or non-stationary signals. Wavelets can be used as building blocks in expansion of signals similar to Fourier series which uses the wave (sinusoid) to represent signals or functions.

4.2.2 Continuous Wavelet Transforms

Wavelets are functions generated from one single function (basis function) called the *prototype* or *mother wavelet* by *dilations* (scalings) and *translations* (shifts) in time (frequency) domain.

$$\psi_{a,b}(t) = \frac{1}{\sqrt{a}} \psi\left(\frac{t-b}{a}\right), \quad \text{where} \quad a, b \in \Re (a > 0) \tag{4.1}$$

Parameter a is a scaling factor, and b is shifting factor. Normalization ensures that $||\Psi_{b,a}(t)|| = ||\Psi(t)||$. The mother wavelet has to satisfy the following admissibility condition:

$$C_\psi = \int_{-\infty}^{\infty} \frac{|\Psi(\omega)|^2}{\omega} d\omega < \infty, \tag{4.2}$$

where $\Psi(\omega)$ is the Fourier transform of $\Psi(t)$.

In practice, $\Psi(\omega)$ will have sufficient decay, so that the admissibility condition reduces to

$$\int_{-\infty}^{\infty} \Psi(t) dt = \Psi(0) = 0 \tag{4.3}$$

Thus, the wavelet will have band-pass behavior.

The 'continuous wavelet transform' (CoWT) of a function $f(t) \in \Re$ is defined as:

$$\text{CoWT}_f(a,b) = <\Psi_{a,b}(t), f(t) >= \frac{1}{\sqrt{a}} \int_{-\infty}^{\infty} f(t) \Psi_{b,a}^* \left(\frac{t-b}{a}\right) dt \tag{4.4}$$

where a and b are two arbitrary real numbers. The variables a and b represent the parameters for *dilations* and *translations*, respectively, in the time axis. The parameter a causes contraction of $\psi(t)$ in the time axis, when $a < 1$, and expansion or stretching, when $a > 1$. The normalizing factor $1/\sqrt{|a|}$ ensures that energy stays the same for all a and b.

4.2.3 Discrete Time Wavelet Transforms

For discrete wavelet transform, the dilation and translation parameters are discretized using

$$a = a_0^{-j}, \quad b = kb_0 a_0^{-j} \tag{4.5}$$

where j and k are integers. Substituting a and b in Eq. (4.4), the discrete wavelets can be represented by:

$$\psi_{j,k}(t) = a_0^{j/2} \psi\left(a_0^j t - kb_0\right) \tag{4.6}$$

So, the wavelet coefficients for function $f(t)$ are given by:

$$c_{j,k}(f) = a_0^{j/2} \int f(t) \psi\left(a_0^j t - kb_0\right) dt \tag{4.7}$$

The most common choice here is $a_0 = 2$ and $b_0 = 1$; hence, $a = 2^j$ and $b = k2^j$. This way of sampling is known as *dyadic sampling*, and the corresponding decomposition of the signals is called the *dyadic decomposition*.

So, for dyadic decomposition, the wavelet coefficients can be derived as:

$$c_{j,k}(f) = 2^{j/2} \int f(t) \psi(2^j t - k) dt \qquad (4.8)$$

The signal $f(t)$ can be reconstructed from the discrete wavelet coefficients as:

$$f(t) = \sum_{j=-\infty}^{\infty} \sum_{k=-\infty}^{\infty} c_{j,k}(f) \psi_{j,k}(t) \qquad (4.9)$$

The transform shown in Eq. (4.8) is often called the discrete time wavelet transform (DTWT).

4.2.4 Discrete Wavelet Transform

The DWT can be explained using *Mallat's* multi-resolution representation of signals [20].

4.2.4.1 Concept of Multi-resolution Analysis

Multi-resolution analysis (MRA) involves two sets of functions: scaling function and wavelet function.

Scaling function

A set of scaling functions is defined in terms of the basic scaling function as:

$$\phi_k(t) = \phi(t - k) \quad k \in Z \quad \phi \in L^2 \qquad (4.10)$$

A family of scaling functions can be generated by scaling and translating the basic scaling function in dyadic fashion,

$$\phi_{j,k}(t) = 2^{j/2} \phi\left(2^j t - k\right), \quad \text{for all integers } k \in Z \qquad (4.11)$$

The subspace of $L^2(R)^1$ spanned by these functions is defined as:

$$V_j = \overline{\text{Span}\{\phi_k(2^j t)\}} = \overline{\text{Span}\{\phi_{j,k}(t)\}} \qquad (4.12)$$

Any function $f(t) \in V_0$ can be expressed as:

$$f(t) = \sum_k a_k \phi\left(2^j t + k\right) \qquad (4.13)$$

---

[1] L indicates a Lebesgue integral, '2' denotes the integral of the square of the modulus of the function, and R states that the independent variable of integration t is a number over the whole real line.

Fig. 4.2 Nesting of spanned
scaling and wavelet spaces

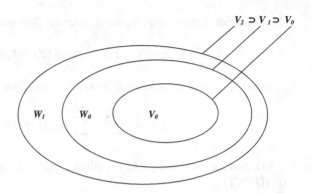

For $j > 0$, $\phi_{j,k}(t)$ represents finer detail, whereas for $j < 0$, it represents coarser information. Consider the nesting of spanned spaces as, $V_j \subset V_{j+1}$ for all $j \in Z$, with $V_{-\infty} = \{0\}$ and $V_\infty = L^2$ as shown in Fig. 4.2.

Then, the higher-resolution signals will also contain lower-resolution signals. So, the scaling function can be expressed as:

$$\phi(t) = \sqrt{2} \sum_n h_0(n)\phi(2t - n) \tag{4.14}$$

where the coefficients $h_0(n)$ are sequence of real or complex numbers known as *scaling function coefficients (or scaling filter)*. The factor $\sqrt{2}$ maintains the norm of the scaling function with the scale of 2.

Wavelet function

A set of wavelet functions is spanned by the differences between the spaces of scaling function as:

$$V_{j+1} = V_j \oplus W_j \tag{4.15}$$

where W_j is the orthogonal complement of V_j in V_{j+1}. As these wavelets also reside in the space spanned by the next resolution scaling function, they can be represented as:

$$\psi(t) = \sqrt{2} \sum_n h_1(n)\phi(2t - n) \tag{4.16}$$

where the coefficients $h_1(n)$ are sequence of real or complex numbers known as *wavelet function coefficients*.

4.3 Implementation of DWT

It has been proved that using MRA, the DWT can be expressed in terms of FIR filters using *Mallat's pyramid algorithm* [20–22]. The input signal is filtered in parallel by a low-pass filter and a high-pass filter to give approximation (coarser)

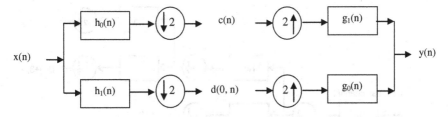

Fig. 4.3 Analysis and synthesis FB for implementing DWT

and details of the input signal as shown in Fig. 4.3. The approximation part can be further decomposed up to level j. This forms the *analysis* (forward) *FB*. During reconstruction, the *synthesis* (inverse) FB is used.

MRA is supported by two functions of DWT: the first one is called as 'scaling function $[\varphi(t)]$' and second one as 'wavelet function $[\Psi(t)]$.'

The scaling function satisfies the following 2-scale (dilation or refinement) equation,

$$\varphi(t) = \sqrt{2} \sum_{n=-\infty}^{\infty} h_0[n]\varphi(2t - n), \quad n \in Z \tag{4.17a}$$

where it satisfies the following admissibility condition:

$$\sum_{n} h_0[n] = \sqrt{2} \tag{4.17b}$$

The wavelet function satisfies the similar equation:

$$\Psi(t) = \sqrt{2} \sum_{n=-\infty}^{\infty} h_1[n]\varphi(2t - n), \quad n \in Z \tag{4.18a}$$

with the following conditions:

$$\sum_{n} h_1[n] = 0 \tag{4.18b}$$

and

$$h_1[n] = (-1)^n h_0[-n + 1] \tag{4.19}$$

Thus, $h_0[n]$ and $h_1[n]$ can be viewed as impulse response of low-pass and high-pass filters, respectively.

The successive lower-resolution coefficients are then recursively computed using Eq. (4.20a) and Eq. (4.20b):

$$C_f(j + 1, k) = \sum_{n} h_0[n - 2k]C_f(j, n) \tag{4.20a}$$

$$D_f(j + 1, k) = \sum_{n} h_1[n - 2k]D_f(j, n) \tag{4.20b}$$

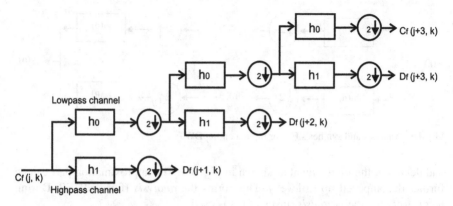

Fig. 4.4 Two-channel, three-level analysis filter bank with 1D DWT

These equations can be implemented as a tree-structured filter bank shown in Fig. 4.4 [20].

Because of the orthonormal wavelet basis, this 3-level analysis filter bank also satisfies the synthesis of high-resolution scaling coefficients from the next immediate level lower-resolution scaling and wavelet coefficients as:

$$C_f(j,k) = \sum_n h_0[k-2n]C_f(j+1,n) + \sum_n h_1[k-2n]D_f(j+1,n) \quad (4.21)$$

The sparse representation with energy compaction makes the standard DWT widely accepted for signal compression. The reconstruction filter bank structure follows the recursive synthesis with reconstruction filters $[\widetilde{h_0}]$ *and* $[\widetilde{h_1}]$ which are identical to their corresponding decomposition filters and but with time reversal.

4.3.1 Perfect Reconstruction

The most important criterion with filter bank implementation is perfect reconstruction [23–27]. For perfect reconstruction analysis and synthesis, filters need to be either the orthogonal wavelet bases or biorthogonal wavelet. The condition for perfect reconstruction (PR) is given by Eq. (4.22a): A simple 2-channel filter bank model is shown in Fig. 4.3

$$H_0(z)\widetilde{H}_0(z) + H_1(z)\widetilde{H}_1(Z) = 2 \qquad (4.22a)$$

$$H_0(z)\widetilde{H}_0(-z) + H_1(z)\widetilde{H}_1(-z) = 0 \qquad (4.22b)$$

where $H_0(Z)$, $H_1(Z)$, $\widetilde{H}_0(z)$, and $\widetilde{H}_1(z)$ are the Z-transforms of $h_0[n]$, $h_1[n]$, $\widetilde{h}_0[n]$, and $\widetilde{h}_1[n]$, respectively.

Daubechies [28] introduced a finite support orthonormal wavelet ψ and the associated filter bank that can be realized through finite-tap FIR filters. Orthogonal and Biorthogonal Wavelets regulate [26, 29–31] the measure of smoothness of

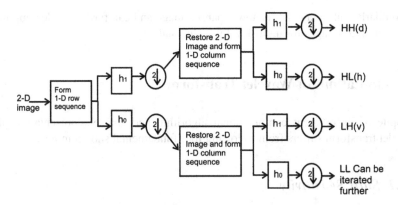

Fig. 4.5 Single-level analysis filter bank for 2D DWT

mother wavelet. The necessary condition for smoothness (regularity) is to have at least one zero at $\omega = \pi$ in low-pass filter (scaling function).

4.3.2 Two-Dimensional Discrete Wavelet Transform

An image is a 2D signal; the DWT decomposes an image $f(x, y) \in L^2(\mathfrak{R}^2)$ in terms of a set of shifted and dilated wavelet functions $\{\psi^{0°}, \psi^{90°}, \psi^{\pm 45°}\}$ and a scaling function $\phi(x, y)$

$$f(x, y) = \sum_{k \in Z^2} C_{fJ_0, k} \varphi_{J_0, k}(x, y) + \sum_{j \in \theta} \sum_{j \geq J_0} \sum_{k \in Z^2} D_{fj, k}^b \psi_{j, k}^b(x, y) \qquad (4.23a)$$

with

$$\varphi_{J_0, k}(x, y) = 2^{J_0} \varphi \left(2^{J_0}(x, y) - k \right) \qquad (4.23b)$$

$$\psi_{j, k}^b(x, y) = 2^j \psi^b \left(2^j(x, y) - k \right) \qquad (4.23c)$$

and

$$b \in \theta = \{0°, 90°, \pm 44°\} \qquad (4.23d)$$

Thus, the filter bank implementation of standard DWT for images is viewed as 2D DWT which is only an extension of 1D DWT applied separately on rows and columns of an image. The 2D DWT implementation of an analysis filter bank structure produces three detailed sub-images (HL, LH, and HH) corresponding to three different directional orientations (horizontal, vertical, and diagonal) and a lower-resolution sub-image LL as shown in Fig. 4.5.

Similarly, sub-image LL is now a parent image and can further be decomposed into four child images for multi-level wavelet analysis.

4.4 Limitations of Wavelet Transforms

In spite of its efficient computational algorithm and sparse representation, the wavelet transform suffers from following four fundamental shortcomings.

4.4.1 Shift Sensitivity

DWT is shift sensitive. A small shift/small shifts in the input signal can result in large differences in DWT coefficients at different scales. The shift sensitivity arises from down samplers in the DWT implementation [32]. Shift sensitivity is a highly undesirable property in pattern recognition application because it implies that same two patterns with small spatial shifts will produce widely different feature vectors.

4.4.2 Poor Directionality

Two-dimensional DWT can resolve only three spatial-domain feature orientations: horizontal (HL), vertical (LH), and diagonal (HH). Natural images (irises) contain a number of smooth regions and edges with random orientations; hence, poor directionality affects the optimal representation of natural images with the help of separable standard 2D DWT.

4.4.3 Absence of Phase Information

DWT implementations use filtering with real coefficient filters, associated with real wavelets, resulting in real-valued approximations and detail, which cannot provide the local phase information.

Edges and other singularities in signal processing applications manifest themselves as oscillating coefficients in the wavelet domain. The amplitude of these coefficients describes the strength of the singularity, while the phase indicates the location of singularity. In order to determine the correct value of localized envelope, phase of an oscillating function is essential along with the magnitude.

4.4.4 Aliasing

The wide spacing of the wavelet coefficient samples, or equivalently the fact that the wavelet coefficients are computed via iterated discrete-time downsampling operations interspersed with non-ideal low-pass and high-pass filters, results in substantial aliasing. Therefore, DWT is not suitable for the analysis of high-frequency signals with relatively narrow bandwidth.

4.5 Hilbert Transform and Analytic Signal

The key is to note that the Fourier transform does not suffer from these problems because Fourier transform is complex. Thus, the prime cause of above-stated problems of DWT is its real-valued coefficients. And all the above-stated limitations can be overcome if coefficients can be made complex-valued by finding the imaginary part. This is conceptually possible with the help of Hilbert transform.

Mathematically, a complex sinusoid $Ae^{j(\omega t+\theta)}$ is simpler than a real sinusoid $A\cos(\omega t + \theta)$ because $e^{j(\omega t+\theta)}$ consists of one frequency ω, while $\cos(\omega t + \theta)$ consists of two frequencies ω and $-\omega$. This is true for all real signals. Also, one of the basic problems in many signal processing applications is about retrieval of instantaneous amplitude A and frequency ω of a real-modulated signal $x(t) = A\cos(\omega t + \theta)$. Retrieval of A becomes ill-posed when $\cos(\omega t + \theta) \approx 0$. The retrieval of instantaneous amplitude and instantaneous frequency is well-posed, simple, and straightforward[2] in case of complex sinusoid. So, it is often preferred to convert real sinusoids into complex sinusoids before processing them further.

A real sinusoid, $A\cos(\omega t + \theta)$, can be converted into a complex sinusoid by generating a phase quadrature component, $A\sin(\omega t + \theta)$ to serve as the *imaginary part*. This phase quadrature component can be generated from the in-phase component by a simple quarter-cycle time shift. Complicated signals, which can be expressed as a sum of sinusoids, can be passed through a filter which shifts each sinusoidal component by a quarter cycles. This filter is called as a *Hilbert transform filter*. This filter has unit magnitude at all frequencies and introduces a phase shift of $-\pi/2$ at each positive frequency and $+\pi/2$ at each negative frequency. When a real signal $h(t)$ and its Hilbert transform $g(t)$ are used to form a new complex signal, $f(t) = h(t) + jg(t)$, the signal $f(t)$ is called the (complex) *analytic signal* corresponding to the real signal $h(t)$. This analytic signal does not have any negative-frequency components, i.e., the spectrum of analytic signal is one-sided as shown in Fig. 4.6.

This quadrature representation of real function with the help of Hilbert transform provides non-negative spectral representation in Fourier domain [33], which saves half the bandwidth which is helpful to avoid aliasing of filter bands. Reduced aliasing of filter bands is the key for shift-invariant property of CWT.

---

[2] Instantaneous peak amplitude at any time can be computed by taking square root of the sum of the squares of the real and imaginary parts. Instantaneous frequency can be obtained by differentiating the phase of the complex sinusoid.

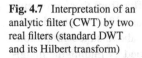

Fig. 4.6 Magnitude spectrum of **a** original signal h(t) and **b** analytic signal f(t)

Fig. 4.7 Interpretation of an
analytic filter (CWT) by two
real filters (standard DWT
and its Hilbert transform)

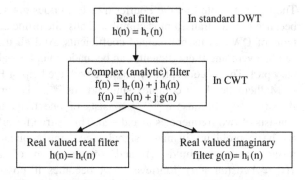

The same concept of quadrature formulation is extended to the filter bank structure of standard DWT to obtain complex wavelet transform having standard DWT and Hilbert transform of standard DWT as its real part and imaginary part, respectively, as illustrated in Fig. 4.7.

4.6 Complex Wavelet Transform

Consider a complex scaling function and a complex wavelet,

$$\phi_c(t) = \phi_r(t) + j\phi_i(t)$$
$$\psi_c(t) = \psi_r(t) + j\psi_i(t) \tag{4.24}$$

where the subscripts r and i indicate real and imaginary components of corresponding scaling and wavelet functions such that

$$\phi_i(t) \approx H\{\phi_r(t)\}$$
$$\psi_i(t) \approx H\{\psi_r(t)\} \tag{4.25}$$

Here, $H(\cdot)$ indicates Hilbert transform.

As the Hilbert transform is infinitely extended in both time and frequency domains (global in nature), it cannot be directly applied to FIR wavelet filter bank [and the Hilbert pair filters will be of infinite length (IIR filters)].

Filters are also required to be orthogonal, symmetric, and linear phase. The orthogonality is necessary to preserve the energy in transform domain. The

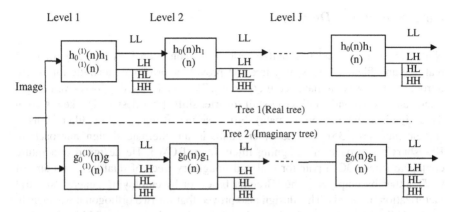

Fig. 4.8   Analysis FB for dual-tree CWT

symmetry property of filter makes it easy to handle the boundary problem for finite-length signals [32]. Linear phase response of the filter is necessary, to reduce the visually objectionable artifacts caused by nonlinear phase distortion, for the quality of image [34]. However, impulse responses of orthogonal, symmetric, and linear phase filters lack quadrature between real and imaginary parts and therefore cannot form a Hilbert pair. Therefore, the development of an invertible analytic (complex) wavelet transform is not as straightforward as expected.

Nick Kingsbury and Selesnick introduced an effective approach for the implementation of complex wavelet transform called as dual-tree CWT with two almost similar versions. These CWTs employ two conventional DWT filter bank trees working in parallel such that the respective filters of both the trees are in approximate quadrature. The filter bank structure of both dual-tree complex wavelet transforms (DT-DWTs) is same, but the design methods to generate the filter coefficients are different. Both DT-DWTs provide phase information; they are shift invariant with improved directionality [16, 17, 35]. In this work, we followed the Selesnick approach.

4.6.1 The Dual-Tree Approach for Complex Wavelets

This approach employs two real wavelet trees; the first (upper) gives the real part, while the second (lower) gives the imaginary part of the CWT as shown in Fig. 4.8. These trees are themselves real and use two different sets of PR filters. However, they are designed such that the overall transform is analytic.

Let $h_0(n)$ and $h_1(n)$ be the low-pass and high-pass filter pairs for the upper filter bank and $g_0(n)$ and $g_1(n)$ be the low-pass and high-pass filter pairs for the lower filter bank. The filters used for first stage should be different from the remaining stages. Failing this prevents the frequency response to be one-sided (analytic) [15].

4.6.2 Selesnick's Dual Tree

Nink Kingsbury proved that half-sample delay condition [17] between filters of real tree and filters of imaginary tree can provide approximate analytic filters. He introduced two ways to have required delays. The first is based on odd–even-length filters, and the second employs Q-shift (quarter shift) filter design. The key issue in the design is to obtain (approximate) shift invariance using any of the filter forms.

Selesnick [16, 35], on the other hand, is an alternate design approach to Kingsbury's approach of designing filters for DT-CWT. He designed quadrature conjugate filter (QCF) pair for real and imaginary trees of dual-tree structure of CWT either by employing the 'Grobner bases' or by employing general 'spectral factorization' methods. This design also proved that for two orthogonal wavelets to form a Hilbert transform pair, the scaling filters of both trees should be offset by a half sample.

The methodology of Selesnick DT-CWT filter design for the filter bank structure as shown in Fig. 4.10 is briefly outlined here.

Let $h_0(n)$ be the low-pass filter and $h_1(n)$ be the corresponding high-pass filter obtained from $h_0(n)$ to form QCF pair of real tree [36] such that

$$h_1(n) = (-1)^{(1-n)} h_0(1-n) \qquad (4.26a)$$

using Z-transform, it can be represented as:

$$H_0(Z)H_0\left(\frac{1}{Z}\right) + H_0(-Z)H_0\left(-\frac{1}{Z}\right) = 2$$
$$H_1(Z) = \frac{1}{z}H_0\left(-\frac{1}{Z}\right) \qquad (4.26b)$$

Similarly, second QCF pair is formed by low-pass filter, $g_0(n)$, and high-pass filter, $g_1(n)$, of imaginary tree of DT-CWT, such that $g_0(n) \cong h_0(n-0.5)$

$$g_1(n) = (-1)^{(1-n)}g_0(1-n) \qquad (4.27a)$$

Using Z-transform, it can be represented as:

$$G_0(Z)H_0\left(\frac{1}{Z}\right) + G_0(-Z)G_0\left(-\frac{1}{Z}\right) = 2$$
$$G_1(Z) = \frac{1}{z}G_0\left(-\frac{1}{Z}\right) \qquad (4.27b)$$

For all these real-valued filters, $\{h_0, h_1\}$ and $\{g_0, g_1\}$, of DT-CWT, the dilation and wavelet equations give the scaling and wavelet functions for the real tree and imaginary tree given by Eqs. (4.28a) and (4.29a), respectively:

$$\varphi_h(t) = \sqrt{2}\sum_n h_0(n)\varphi_h(2t-n) \qquad (4.28a)$$

$$\Psi_h(t) = \sqrt{2}\sum_n h_1(n)\Psi_h(2t-n) \qquad (4.28b)$$

$$\varphi_g(t) = \sqrt{2} \sum_n g_0(n) \varphi_g(2t - n) \tag{4.29a}$$

$$\Psi_g(t) = \sqrt{2} \sum_n g_1(n) \Psi_g(2t - n) \tag{4.29b}$$

As discussed in Sect. 4.5, $\Psi_g(t)$ is the Hilbert transform of $\Psi_g(t)$, $\Psi_g(t) = H\{\Psi_h(t)\}$, if

$$\Psi_g(\omega) = -j\Psi_h(\omega), \quad \omega > 0 \tag{4.30a}$$

$$= j\Psi_h(\omega), \quad \omega < 0 \tag{4.30b}$$

where $\Psi_h(\omega)$ and $\Psi_g(\omega)$ are the Fourier transform of the wavelet functions (high-pass filters) of real and imaginary parts of DT-CWT.

Similarly, $H_0(\omega) = \text{DTFT}\{h_0(n)\}$, $G_0(\omega) = \text{DTFT}\{g_0(n)\}$ are the Fourier transform of scaling functions (low-pass filters), h_0 and g_0, respectively.

If $\Psi_h(t)$ and $\Psi_g(t)$ form the Hilbert transform pair, then

$$|\Psi_h(\omega)| = |\Psi_g(\omega)| \quad \text{and therefore} \quad |H_0(\omega)| = |G_0(\omega)| \tag{4.31}$$

Thus, low-pass filters of real tree and imaginary tree maintain the following relation:

$$G_0(\omega) = H_0(\omega)e^{-j(\omega/2)}, \quad |\omega| < \pi \tag{4.32}$$

In time domain, the digital filter $g_0(n)$ is a half-sample delayed version of $h_0(n)$,

$$g_0(n) = h_0\left(n - \frac{1}{2}\right) \tag{4.33}$$

As a half-sample delay cannot be implemented with FIR filter, the approximate design of FIR filters $h_0(n)$ and $g_0(n)$ is employed with Grobner bases or spectral factorization [16, 35]. Orthogonal filter sets are used, and coefficients of all the filters are obtained from the low-pass filter of real tree, $h_0(n)$. All the synthesis filters are the time-reversed version of their corresponding analysis filters.

4.6.3 2D DT-CWT

The 2D DT-DWT structure is an extension of conjugate filtering in 2D case. The pairs of conjugate filters are applied to two directions (x and y), which can be expressed by Eq. (4.32):

$$(h_x + j g_x)(h_y + j g_y) = (h_x h_y - g_x g_y) + j (h_x g_y + g_x h_y) \tag{4.34}$$

where h_x, g_x, h_y, and g_y are the filter sets associated with scaling and wavelet functions ($\{\varphi_h(x), \psi_h(x)\}$, $\{\varphi_g(x), \psi_g(x)\}$, $\{\varphi_h(y), \psi_h(y)\}$, $\{\varphi_g(y), \psi_g(y)\}$) of real tree (tree-a) and imaginary tree (tree-b) operated in x-direction (row-wise implementation) and y-direction (column-wise implementation), respectively.

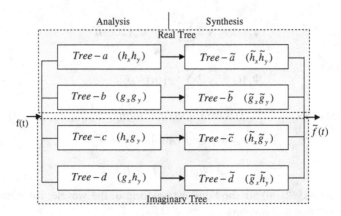

Fig. 4.9 filter bank structure for 2D DT-DWT

Approximate analytic wavelet functions associated with x-direction and y-direction are given by Eq. (4.35a):

$$\psi(x) = \psi_h(x) + j\psi_g(x) \tag{4.35a}$$

and

$$\psi(y) = \psi_h(y) + j\psi_g(y) \tag{4.35b}$$

Similarly, approximate analytic scaling functions associated with x-direction and y-direction are given by Eq. (4.36a):

$$\varphi(x) = \varphi_h(x) + j\varphi_g(x) \tag{4.36a}$$

and

$$\varphi(y) = \varphi_h(y) + j\varphi_g(y) \tag{4.36b}$$

From Eq. (4.34), it is clear that DT-CWT structure needs four trees for its implementation as shown in Fig. 4.9.

Each of the four trees is implemented by two filter banks of real and imaginary trees operated row-wise and column-wise so that Eq. (4.34) is implemented. The filter bank structures of tree-a, tree-b, tree-c, and tree-d are shown in Fig. 4.10. Real part of 2D DT-CWT is formed by tree-a and tree-b, and the tree-c and tree-d form the imaginary part of 2D DT-CWT.

If we define the separable 2D wavelet bases as

$$\begin{aligned}
\psi_{1,1}(x,y) &= \phi_h(x)\psi_h(y), & \psi_{2,1}(x,y) &= \phi_g(x)\psi_g(y)\\
\psi_{1,2}(x,y) &= \psi_h(x)\phi_h(y), & \psi_{2,2}(x,y) &= \psi_g(x)\phi_g(y)\\
\psi_{1,3}(x,y) &= \psi_h(x)\psi_h(y), & \psi_{2,3}(x,y) &= \psi_g(x)\psi_g(y)
\end{aligned} \tag{4.37a}$$

then the wavelets are defined by:

$$\psi_i(x,y) = \frac{1}{\sqrt{2}}\left(\psi_{1,i}(x,y) + \psi_{2,i}(x,y)\right)$$

$$\psi_{i+3}(x,y) = \frac{1}{\sqrt{2}}\left(\psi_{1,i}(x,y) - \psi_{2,i}(x,y)\right) \tag{4.37b}$$

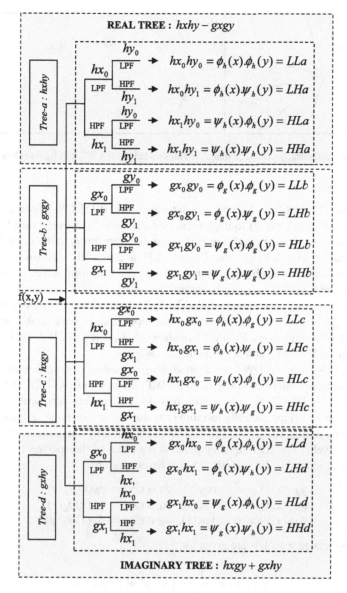

Fig. 4.10 Details of filter bank structure of real tree and imaginary tree of DT-CWT

for $i = 1, 2, 3$. The normalization $1/\sqrt{2}$ is used so that sum/difference operation constitutes an orthonormal operation. These wavelets are strongly oriented at angles of $\pm 15°$, $\pm 45°$, and $\pm 75°$, and this is *non-redundant approach* and also termed as *real dual-tree DWT(R-DT-CWT)*. This approach seeks $\psi_c(t)$ that forms an orthonormal or biorthogonal basis but fails in addressing in the limitations of DWT in totality.

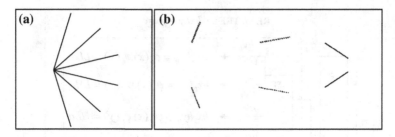

Fig. 4.11 **a** Synthetically generated image. **b** Thresholded R-DT-CWT sub-bands at level 3

The real 2D dual-tree discrete wavelet transform (R-DT-CWT) can be implemented using $\{h_0(n), h_1(n)\}$ to implement one separable 2D wavelet transform and $\{g_0(n), g_1(n)\}$ to another. Applying both separable transforms to the same 2D data gives a total of six sub-bands: two HL, two LH, and two HH sub-bands. Take the sum and difference of each pair of sub-bands to get a transform which is two times expansive and free of the checkerboard artifact.

An implemented illustration of R-DT-CWT operated on a synthetically generated image is as shown in Fig. 4.11a. The input images have spokes in $\pm 15°$, $\pm 45°$, and $\pm 75°$ directions. Figure 4.11b shows that the R-DT-CWT sub-bands are oriented at $\pm 15°$, $\pm 45°$, and $\pm 75°$ angles.

All the limitations of DWT are addressed by *redundant approach* which is also called as complex dual-tree CWT (C-DT-CWT). In this book, C-DT-CWT is simply denoted by DT-CWT. This approach seeks a redundant representation, with both $\psi_r(t)$ and $\psi_i(t)$ individually forming orthonormal or biorthogonal basis.

Similarly, in Eqs. (4.32) and (4.33), if we consider only the complex part of the complex wavelet, the 2D frequency plane is the same as the spectrum of the real part with the separable 2D wavelet [15–19] bases given by:

$$\begin{aligned}
\psi_{3,1}(x,y) &= \phi_g(x)\psi_h(y), \quad \psi_{4,1}(x,y) = \phi_h(x)\psi_g(y) \\
\psi_{3,2}(x,y) &= \psi_g(x)\phi_h(y), \quad \psi_{4,2}(x,y) = \psi_h(x)\phi_g(y) \\
\psi_{3,3}(x,y) &= \psi_g(x)\psi_h(y), \quad \psi_{4,3}(x,y) = \psi_h(x)\psi_g(y)
\end{aligned} \qquad (4.38)$$

and the associated wavelets are given by:

$$\begin{aligned}
\psi_i(x,y) &= \frac{1}{\sqrt{2}}(\psi_{3,i}(x,y) + \psi_{4,i}(x,y)) \\
\psi_{i+3}(x,y) &= \frac{1}{\sqrt{2}}(\psi_{3,i}(x,y) - \psi_{4,i}(x,y))
\end{aligned} \qquad (4.39)$$

for $i = 1, 2, 3$.

When we combine the two parts of the dual tree into a complex basis function (and its conjugate), then we also separate positive frequencies from negative frequencies. The real and imaginary parts of each complex wavelet are oriented at the same angle, and the magnitude of each complex wavelet is an approximately circular bell-shaped function. Figure 4.12 illustrates a set of six oriented complex wavelets obtained in this way.

4.6.4 2D DT-CWT—Wavelet Filter Design

Let $\phi_h(t)$ and $\psi_h(t)$ be the scaling and wavelet functions associated with real tree and $\phi_g(t)$ and $\psi_g(t)$ be the scaling and wavelet functions associated with imaginary tree. Then, wavelets associated with 2D DT-CWT are given by the following equations [15–17], which are used for implementation of four trees of DT-CWT as shown in Fig. 4.10.

$$
\begin{aligned}
\phi(x)\psi(y) &= [\phi_h(x) + j\phi_g(x)][\psi_h(y) + j\psi_g(y)] \\
&= \phi_h(x)\psi_h(y) - \phi_g(x)\psi_g(y) + j\phi_g(x)\psi_h(y) + j\phi_h(x)\psi_g(y) \\
\phi(x)\overline{\psi(y)} &= [\phi_h(x) + j\phi_g(x)][\psi_h(y) + j\psi_g(y)] \\
&= \phi_h(x)\psi_h(y) + \phi_g(x)\psi_g(y) + j\phi_g(x)\psi_h(y) - j\phi_h(x)\psi_g(y) \\
\overline{\phi(x)}\psi(y) &= [\phi_h(x) - j\phi_g(x)][\psi_h(y) + j\psi_g(y)] \\
&= \phi_h(x)\psi_h(y) + \phi_g(x)\psi_g(y) - j\phi_g(x)\psi_h(y) + j\phi_h(x)\psi_g(y) \\
\overline{\phi(x)}\overline{\psi(y)} &= [\phi_h(x) - j\phi_g(x)][\psi_h(y) + j\psi_g(y)] \\
&= \phi_h(x)\psi_h(y) - \phi_g(x)\psi_g(y) - j\phi_g(x)\psi_h(y) + j\phi_h(x)\psi_g(y) \\
\psi(x)\phi(y) &= [\psi_h(x) + j\psi_g(x)][\phi_h(y) + j\phi_g(y)] \\
&= \psi_h(x)\phi_h(y) - \psi_g(x)\phi_g(y) + j\psi_g(x)\phi_h(y) + j\psi_h(x)\phi_g(y) \\
\psi(x)\overline{\phi(y)} &= [\psi_h(x) + j\psi_g(x)][\phi_h(y) + j\phi_g(y)] \\
&= \psi_h(x)\phi_h(y) + \psi_g(x)\phi_g(y) + j\psi_g(x)\phi_h(y) - j\psi_h(x)\phi_g(y) \\
\overline{\psi(x)}\phi(y) &= [\psi_h(x) - j\psi_g(x)][\phi_h(y) + j\phi_g(y)] \\
&= \psi_h(x)\phi_h(y) + \psi_g(x)\phi_g(y) - j\psi_g(x)\phi_h(y) + j\psi_h(x)\phi_g(y) \\
\overline{\psi(x)}\overline{\phi(y)} &= [\psi_h(x) - j\psi_g(x)][\phi_h(y) - j\phi_g(y)] \\
&= \psi_h(x)\phi_h(y) - \psi_g(x)\phi_g(y) - j\psi_g(x)\phi_h(y) - j\psi_h(x)\phi_g(y) \\
\psi(x)\psi(y) &= [\psi_h(x) - j\psi_g(x)][\psi_h(y) - j\psi_g(y)] \\
&= \psi_h(x)\psi_h(y) + \psi_g(x)\psi_g(y) + j\psi_g(x)\psi_h(y) - j\psi_h(x)\psi_g(y) \\
\overline{\psi(x)}\psi(y) &= [\psi_h(x) - j\psi_g(x)][\psi_h(y) + j\psi_g(y)] \\
&= \psi_h(x)\psi_h(y) + \psi_g(x)\psi_g(y) - j\psi_g(x)\psi_h(y) + j\psi_h(x)\psi_g(y) \\
\overline{\psi(x)}\overline{\psi(y)} &= [\psi_h(x) - j\psi_g(x)][\psi_h(y) - j\psi_g(y)] \\
&= \psi_h(x)\psi_h(y) - \psi_g(x)\psi_g(y) - j\psi_g(x)\psi_h(y) - j\psi_h(x)\psi_g(y)
\end{aligned}
$$

$$(4.40a\text{–}1)$$

These complex signals have single-sided frequency spectrum. So the resultant wavelets are also one-sided as shown in Fig. 4.13.

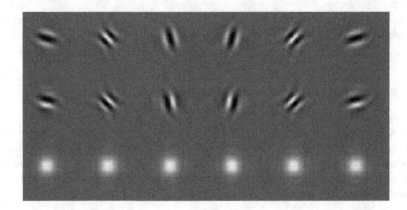

Fig. 4.12 Impulse responses for 2D C-DT-DWT; *first row* is interpreted as the real part and the *second row* as imaginary part of the complex wavelet. The *third row* shows the magnitude response (same for both, real and complex wavelets)

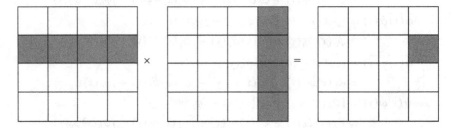

Fig. 4.13 Idealized 2D Fourier spectrums [15]

4.7 DT-CWT Filters—Design and Implementation

Implementation of DT-CWT requires that filter at first stage is different than that at successive stages of wavelet decomposition [15, 17].

If same PR filters are used for each stage, then final stages will not be (approximately) analytic, i.e., $\psi_g(n) \neq H[\psi_h(n)]$ where $H[\psi_h(n)]$ is Hilbert transform of $\psi(n)$. $\psi_g(n) \approx H[\psi_h(n)]$ is achieved if

$$g_0(n) = h_0(n - 0.5) \tag{4.41a}$$

To understand the design of filters at first stage and other stages, consider half-sample delay condition, i.e., $g_0(n) \approx h_0(n - 0.5)$.

That is, scaling function satisfies this half-sample delay condition. That is,

$$\varphi_g(n) \approx \varphi_h(n - 0.5) \tag{4.41b}$$

This implies that the integer translates of $\varphi_g(n)$ fall midway between integer translates of $\varphi_h(n)$, i.e., two scaling functions satisfy interlacing property. Thus, for

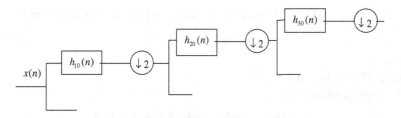

Fig. 4.14 Low-pass filtering at multi-levels

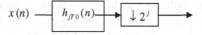

Fig. 4.15 Equivalent representation of cascaded system of Fig. 4.14

DT-CWT to be approximately analytic at each stage, this interlacing property must be followed.

Let us consider that each stage has different PR filters.

$\{h_{j0}, h_{j1}\}$ and $\{g_{j0}, g_{j1}\}$ are the filter bank of real and imaginary trees at jth stage where $j = 1,2,3...L$, for L-level wavelet decomposition. Consider the cumulative low-pass multi-rate filtering [15] as shown in Fig. 4.14.

Thus, an equivalent of cascaded filters [15] at multi-levels can be represented as illustrated in Fig. 4.15. Where $h_{jT0}(n)$ is convolution of impulse responses of filters at each stage, i.e.,

$$h_{jt0} = h_{10}(n) * h_{20}(n) * h_{30}(n) * \cdots * h_{j0}(n) \tag{4.42a}$$

In Z-domain,

$$H_{jT0}(z) = H_{10}(z) \cdot H_{20}(t) \ldots H_{j0}(z) \tag{4.42b}$$

Similarly, for imaginary tree,

$$G_{jT0}(z) = G_{10}(z) \cdot G_{20}(z) \ldots G_{j0}(z) \tag{4.43}$$

To satisfy the interlacing property, the filter at each stage j for $j = 1,2,3...L$ shall be designed in such a way that a translation of $g_{jT}(n)$ by 2^j falls midway between the translates of $h_{jT}(n)$ by 2^j.

Consider the condition at Stage 1 ($j = 1$)

Translates $g_{1T}(n)$ by 2^1 fall midway between the translates at $h_{1T}(n)$ by 2^1.

That is,

$$g_{1T}(n) \approx h_{1T}(n-1) \tag{4.44}$$

Consider the condition at Stage 2 ($j = 2$)

Translates $g_{2T}(n)$ by 2^2 fall midway between the translation at $h_{2T}(n)$ by 2^2.

That is,

$$g_{2T}(n) \approx h_{1T}(n-2) \tag{4.45}$$

Consider the condition at Stage 3 ($j = 3$)

Translates $g_{2T}(n)$ by 2^3 fall midway between the translation at $h_{2T}(n)$ by 2^3

$$g_{3T}(n) \approx h_{1T}(n - 3) \tag{4.46}$$

and so on.

From Figs. 4.14 and 4.15,

At Stage 1, i.e., at $j = 1$

$$h_{1T}(n) = h_{10}(n) \quad \text{and} \quad g_{1T}(n) = g_{10}(n)$$
$$\therefore g_{10}(n) \approx h_{10}(n - 1) \tag{4.47}$$

This is different than half-sample delay condition derived to achieve approximate Hilbert transform pair, and for the first stage, the condition of Eq. (4.43) is satisfied simply by using one sample delayed low-pass filter for imaginary tree. Thus, any set of PR filters is used at first stage of wavelet decomposition.

Now at Stage 2, i.e., $j = 2$,

$$g_{2T}(n) \approx g_{2T}(n - 2)$$

Taking Fourier transform,

$$G_{2T}(\omega) \approx e^{-j2\omega} H_{2T}(\omega) \tag{4.48}$$

But from Figs. 4.14 and 4.15

$$G_{2T}(z) \approx G_{10}(z) \cdot G_{20}(z^2) \tag{4.49a}$$

and

$$H_{2T}(z) = H_{10}(z) \cdot H_{20}(z^2) \tag{4.49b}$$

Therefore, from Eqs. (4.44) and (4.45), $G_{10}(z) \cdot G_{20}(z^2) = z^{-2} H_{10}(z) \cdot G_{20}(z^2)$ at $z = 1 \cdot e^{j\omega}$,

$$G_{10}(e^{j\omega}) \cdot G_{20}(e^{j2\omega}) \approx e^{-j2\omega} H_{10}(e^{j\omega}) \cdot H_{20}(e^{j2\omega}) \tag{4.50}$$

Substituting Eq. (4.48) in (4.50)

$$G_{20}(e^{j2\omega}) \approx e^{-j\omega} H_{20}(e^{j2\omega})$$

That is,

$$G_{20}(e^{j\omega}) \approx e^{-j\frac{\omega}{2}} H_{20}(e^{j\omega}) \tag{4.51}$$

That is, in time domain,

$$g_{20}(n) \approx h_{20}(n - 0.5) \tag{4.52}$$

Similarly, at 3rd stage, i.e., at $j = 3$

$$G_{3T}(e^{j\omega}) \approx e^{-j4\omega} H_{3T}(e^{j\omega}) \tag{4.53}$$

But,

$$G_{10}(e^{j\omega}) \cdot G_{20}(e^{j2\omega}) \cdot G_{30}(e^{j4\omega}) \approx e^{-j4\omega} H_{10}(e^{j\omega}) \cdot H_{20}(e^{j2\omega}) \cdot H_{30}(e^{j4\omega})$$
$$\tag{4.54}$$

From Eqs. (4.47), (4.51)–(4.53),

$$G_{30}(e^{j4\omega}) \approx e^{-j2\omega} H_{30}(e^{j4\omega})$$
$$G_{30}(e^{j\omega}) \approx e^{-j0.5\omega} H_{30}(e^{j\omega})$$

(4.55a)

In time domain,

$$g_{30}(n) \approx h_{30}(n - 0.5)$$

(4.55b)

Thus, for $j > 1$, (i.e., stages after first stage)

$$g_{j0}(n) \approx h_{j0}(n - 0.5), \quad \text{for} \quad j > 1$$

(4.56)

Therefore, the PR dual-tree filters maintain the half-sample delay condition after first stage.

As per Eq. (4.47), the first-stage filter does not require approximately half-sample delay condition, i.e., it is not required to have approximately analytic filters and they can be any equal-length real-valued PR filters used for DWT.

Thus, we have designed 8-tap Daubechies low-pass filter for first stage of real tree of DT-CWT. The design procedure to obtain the coefficient of this filter is outlined in Sect. 4.7.1.

4.7.1 Design of Low-pass Filter (Scaling Function) of Real Tree of DT-DWT

To obtain the maximum possible numbers of vanishing moments, Daubechies [21, 28] utilized $\frac{N}{2} - 1$ degree of freedom for designing low-pass filter of length N.

We have

$$H(z) = \left(\frac{1 + z^{-1}}{2}\right)^K \cdot Q(z)$$

(4.57)

Here, $K = \frac{N}{2}$

In Fourier domain,

$$H(e^{j\omega}) = \left(\frac{1 + e^{-j\omega}}{2}\right)^K \cdot Q(e^{j\omega})$$

(4.58)

For low-pass filter,

$$|H(\omega)|^2 + |H(\omega + \pi)|^2 = 2$$

(4.59)

Equation (4.58) satisfies condition of LPF as illustrated in Eq. (4.59) if

$$|Q(\omega)|^2 = P(\sin^2(\omega/2))$$

(4.60)

where

$$P(x) = \sum_{C=0}^{C-1} \binom{N-1+C}{C_n} x^C + x^C R\left(\frac{1}{2} - x\right) \tag{4.61}$$

Such that $R(x)$ is an odd polynomial chosen to ensure

$$p(x) \geq 0, \quad \text{for} \quad 0 \leq x \leq 1$$

$h(n)$ is impulse response of low-pass filter of length N
i.e., $h(n) = \{h(0), h(1)\ldots h(N-1)\}$

Taking Z-transform,

$$H(t) = h(0) + h(1)z^{-1} + h(2)z^{-2} + \cdots + h(N-1)z^{-(N-1)} \tag{4.62}$$

Thus, $H(z)$ is $(N-1)$ degree polynomial of z^{-1}.

All the filters of DWT and CWT used in this research for extracting the iris features are designed from the coefficient of low-pass filter. Therefore, design of low-pass filter (scaling function) of wavelet transform is of prime importance. We designed tap-8 filter for stage 1 and tap-10 filter for all stages after stage 1 of wavelet decomposition.

Orthogonality is the most important property of scaling function (low-pass filter) in DWT filter bank implementation, not only for perfect recognition (PR) but also for reducing the redundancy and memory requirements as compared to redundancy and memory requirements in Gabor filter implementation. Moreover, filter must be characterized by the maximum number of vanishing movements for a given support (length of the filter). Daubechies [21, 28] has proposed the methodology to design such type of orthogonal filter that provides maximum number of vanishing movements.

Let $h_0(n)$ be the impulse response of low-pass filter of length N and $H(z)$ be the Z-transform of it having degree of $(N-1)$.

The condition of orthogonality is given by Eq. (4.52)

$$\sum_{n=0}^{n-1} h_0(n) = \sqrt{2} \tag{4.63a}$$

and

$$\sum_{n=0}^{n-1} h_0^2(n) = 1 \tag{4.63b}$$

To obtain the maximum number of vanishing wavelet moments, consider $Q(z)$ the $(\frac{N}{2} - 1)$ degree polynomial of 'z' such that

$$|Q(z)|^2 = |Q(z)| \cdot \left|Q(z^{-1})\right|^2$$

$$\therefore |Q(z)|^2 = \sum_{n=0}^{\frac{N}{2}-1} \binom{\frac{N}{2}-1+n}{C_n} \left[\frac{1}{2} - \frac{1}{4}(z + z^{-1})\right]^n \tag{4.64}$$

where N is the length of the filter.

Equation (4.57) is used to obtain the coefficients of minimal phase, maximal phase, and mixed phase factorization methods as outlined in the following steps:

Step 1:   Find $|Q(z)|^2 = |Q(z)| \cdot |Q(z^{-1})|$ for tap-8 filter, i.e., $N = 8$

Substitute $N = 8$ in Eq. (4.64)

$$\therefore |Q(z)|^2 = |Q(z)| \left|Q(z^{-1})\right| = \sum_{n=0}^{3} (\overset{n+3}{C_n}) \left[\frac{1}{2} - \frac{1}{4}(z + z^{-1}) \right]^n$$

$$= \left[\overset{0+3}{C_0} \right] \left[\frac{1}{2} - \frac{1}{4}(z + z^{-1}) \right]^0 + \left[\overset{1+3}{C_1} \right] \left[\frac{1}{2} - \frac{1}{4}(z + z^{-1}) \right]^1$$

$$+ \left[\overset{2+3}{C_2} \right] \left[\frac{1}{2} - \frac{1}{4}(t + t^{-1}) \right]^2 + \left[\overset{3+3}{C_3} \right] \left[\frac{1}{2} - \frac{1}{4}(z + z^{-1}) \right]^3$$

$$= 1 + 4 \left[\frac{1}{2} - \frac{1}{4}z - \frac{1}{4}z^{-1} \right] + 10 \left[\frac{1}{2} - \frac{1}{16}(z - z^{-1})^2 - \frac{1}{4}(z + z^{-1}) \right].$$

$$+ 20 \left[\frac{1}{8} - \frac{3}{16}(z - z^{-1}) + \frac{3}{32}(z + z^{-1})^2 - \frac{1}{64}(z + z^{-1})^3 \right]$$

$$\tag{4.65}$$

$$|Q(z)|^2 = |Q(z)| \left|Q(z^{-1})\right| = \frac{5}{16}z^3 + \frac{5}{2}z^2 - \frac{131}{16}z + 13 - \frac{131}{16}z^{-2} + \frac{5}{2}z^{-2} + \frac{5}{16}z^{-2} \tag{4.66a}$$

Step 2:   Find factors of

$$|Q(z)|^2 \tag{4.66b}$$

$$\therefore |Q(z)|^2 = 0 \tag{4.67a}$$

As $|Q(z)|^2 = |Q(z)| \cdot |Q(z^{-1})|$, roots of $|Q(z)|^2$ are in pairs of reciprocals.

$$\therefore \frac{5}{16}z^3 + \frac{5}{2}z^2 - \frac{131}{16}z + 13 - \frac{131}{16}z^{-2} + \frac{5}{2}z^{-2} + \frac{5}{16}z^{-3} = 0 \quad (4.67b)$$

$$z = \{3.0407, 2.0311 + j1.7390, 2.0311 - j1.7390, 0.2841$$
$$+ j0.2432, 0.2841 - j0.2432, 0.3289\} \tag{4.68a}$$

Out of these six roots, three roots belong to $Q(z)$ and the remaining three roots belong to $Q(z^{-1})$. These roots are such that it forms three reciprocal pairs as illustrated by Eq. (4.68b).

$$z = \left\{ z_1, z_2, z_3, \frac{1}{z_3}, \frac{1}{z_2}, \frac{1}{z_1} \right\} \tag{4.68b}$$

Fig. 4.16 Typical example of locations of roots of $|Q(z)|^2$

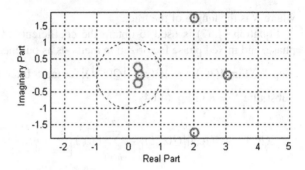

The location of these poles is shown in Fig. 4.16.

$$\therefore Q(z) = [z - 0.3289][z - 0.2841 + j0.2432][z - 0.2841 - j0.2432] \quad (4.69a)$$

$$\therefore Q(z) = z^3 - 0.3289z^2 - 0.13979z + 0.04977 \quad (4.69b)$$

Step 3: Find $H(Z)$ from $Q(Z)$ for $n = (N/2) - 1 = 4$

$$H(z) = \left[\frac{1 + z^{-1}}{2}\right]^4 Q(z) \quad (4.70a)$$

$$\therefore H(Z) = \frac{1}{16}\left(1 + 4Z^{-1} + 6Z^{-2} + 4Z^{-3} + Z^{-4}\right)Q(z) \quad (4.70b)$$

$$\therefore H(Z) = \frac{1}{16}\left(1 + 4Z^{-1} + 6Z^{-2} + 4Z^{-3} + Z^{-4}\right)$$
$$\left(z^3 - 0.3289z^2 - 0.13979z + 0.04977\right) \quad (4.70c)$$

From Eq. (4.70a), we obtain $H(Z)$ of $(N - 1)°$.

Taking inverse Z-transform of Eq. (4.70c) and after normalization, the coefficients of impulse response of length-8 low-pass filter are tabulated in Table 4.1. We used this filter at first stage of wavelet decomposition.

By using the above-stated design procedure, we also obtain the coefficients of impulse response of low-pass filter of length-10 which is used at all states after stage 1 of wavelet decomposition. The coefficients of this filter are tabulated in Table 4.2.

For above-stated filters, the following conditions are satisfied:

$$\sum_{n=0}^{\frac{N}{2}-1} h_0(n) \cdot h_0\left(\frac{N}{2} + n\right) = 0 \quad (4.71)$$

Table 4.1 Length-8 Daubechies low-pass filter for first-level wavelet decomposition

$h(0)$	$h_{10}(1)$	$h_{10}(2)$	$h_{10}(3)$	$h_{10}(4)$	$h_{10}(5)$	$h_{10}(6)$	$h_{10}(7)$
0.2304	0.7148	0.6309	−0.0280	−0.1870	0.0308	0.0329	−0.0106

Table 4.2 Length-10 Daubechies low-pass filter for other than first-level wavelet decomposition

$h_0(0)$	$h_0(1)$	$h_0(2)$	$h_0(3)$	$h_0(4)$
0.160127	0.60392	0.724418	0.138449	−0.24233

$h_0(5)$	$h_0(6)$	$h_0(7)$	$h_0(8)$	$h_0(9)$
0.160127	0.60392	0.724418	0.138449	−0.24233

$$\sum_{n=0}^{N-1} (-1)^n h_0(n) = 0 \tag{4.72}$$

Also,

$$\left. \begin{aligned} |H_0(\omega)|_{\omega=\pi} &= H_0(\pi) = 0 \\ |H_0(\omega)|_{\omega=0} &= H_0(0) = \sqrt{2} \end{aligned} \right\} \tag{4.73}$$

Therefore, these filters are orthogonal low-pass filters.

4.7.2 Design of High-Pass Filter (Wavelet Function) from the Low-Pass Filter

Let Fourier transform of low-pass filter, $h_0(n)$, be $H_0(\omega)$ and corresponding high-pass filter, $h_1(n)$, be $H_1(\omega)$.

Therefore, Fourier transform of high-pass filter is nothing but the Fourier transform of low-pass filter shifted by π.

That is,

$$H_1(\omega) = H_0(\omega + \pi) \tag{4.74}$$

If

$$H_0(\omega) = h_0(0) + h_0(1)e^{-j\omega} + h_0(2)e^{-j2\omega} + \cdots + h_0(N-1)e^{-j(N-1)\omega} \tag{4.75}$$

then

$$H_1(\omega) = h_1(0) - h_1(1)e^{j\omega} + h_1(2)e^{j2\omega} + \cdots + h_1(N-1)e^{j(N-1)\omega} \tag{4.76}$$

where $h_0(n)$ and $h_1(n)$ are even-length filters.

From Eqs. (4.75) and (4.76), it is clear that $h_0(n)$ and $h_1(n)$ form a QCF pair. Therefore, the coefficients of high pass filter [15, 16] can be computed using Eq. (4.77):

$$h_1(n) = (-1)^n h_0(N-1-n) \quad 0 \le n \le N-1 \tag{4.77}$$

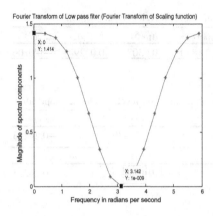
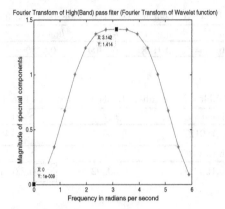

Fig. 4.17   Fourier transform of scaling and wavelet functions of our filter bank

Table 4.3   Coefficients of high-pass filter at stage 1

$h_{11}(0)$	$h_{11}(1)$	$h_{11}(2)$	$h_{11}(3)$	$h_{11}(4)$	$h_{11}(5)$	$h_{11}(6)$	$h_{11}(7)$
−0.0106	−0.0329	0.0308	0.187	−0.028	−0.6309	0.7148	−0.2304

Such that Eqs. (4.73) and (4.74) hold true for $h_1(n)$.
Also,

$$\sum_{n=0}^{N-1} h_1(n) = 0 \qquad (4.78)$$

For above-stated high-pass filter,

$$H_1(0) = 0 \quad \text{and} \quad H_1(\pi) = \sqrt{2} \qquad (4.79)$$

Equation (4.79) confirmed the high-pass filter characteristics of $h_1(n)$.

The Fourier transform of our filters is as shown in Fig. 4.17.

Moreover, the designed low-pass filter and high-pass filter are orthogonal because it satisfies the following Eq. (4.80):

$$\sum_{n=0}^{N-1} h_0(n) \cdot h_1(n) = 0 \qquad (4.80)$$

The coefficients of high-pass filter, $h_{11}(n)$, at all stages after first stage of wavelet decomposition are obtained from $h_{10}(n)$, such that these two filters form a QCF pair. That is,

$$h_{11}(n) = (-1)^n h_{10}(N - 1 - n), \quad 0 \le n \le N - 1 \qquad (4.81)$$

The coefficients of $h_{11}(n)$ obtained from Table 4.1 and Eq. (4.81) are obtained in Table 4.3.

Now, low-pass filter of imaginary tree at first stage, $g_{10}(n)$, is obtained by one sample delay in $h_{11}(n)$.

Table 4.4 Coefficients of $g_{10}(n)$

$g_{10}(0)$	$g_{10}(1)$	$g_{10}(2)$	$g_{10}(3)$	$g_{10}(4)$	$g_{10}(5)$	$g_{10}(6)$	$g_{10}(7)$
0.0329	0.0308	−0.187	−0.028	0.6309	0.7148	0.2304	−0.0106

Table 4.5 Coefficients of $g_{11}(n)$

$g_{11}(0)$	$g_{11}(1)$	$g_{11}(2)$	$g_{11}(3)$	$g_{11}(4)$	$g_{11}(5)$	$g_{11}(6)$	$g_{11}(7)$
−0.0106	−0.2304	0.7148	−0.6309	−0.028	0.187	0.0308	−0.0329

Table 4.6 Coefficients of high-pass filter at stage 1

$h_1(0)$	$h_{11}(1)$	$h_1(2)$	$h_1(3)$	$h_1(4)$
0.003336	0.012583	−0.00624	−0.07758	−0.03225
$h_1(5)$	$h_1(6)$	$h_1(7)$	$h_1(8)$	$h_1(9)$
0.242331	0.138449	−0.72442	0.60392	−0.16013

The coefficients of $g_{10}(n)$ are listed in Table 4.4.

The coefficients of high-pass filter $g_{11}(n)$ are obtained from $g_{10}(n)$, such that these two filters form another QCF pair, and the coefficients of $g_{11}(n)$ obtained with the help of Eq. (4.82) are listed in Table 4.5.

That is,

$$g_{11}(n) = (-1)^n g_{10}(N - 1 - n), \quad 0 \le n \le N - 1 \tag{4.82}$$

4.7.3 Filter Design for Other Stages (After Stage 1) of DT-CWT

Let $h_0(n)$ and $h_1(n)$ be the low-pass filter and high-pass filter of real tree, respectively, and $g_0(n)$ and $g_1(n)$ are the low-pass filter and high-pass filter of imaginary tree, respectively, at all stages after Stage 1 of DT-CWT.

$h_0(n)$ is designed as tap-10 Daubechies low-pass filter as listed in Table 4.2.

The coefficients of high-pass filter of real tree at all stages after first stage, $h_1(n)$, are such that $h_0(n)$ and $h_1(n)$ form a QCF pair. That is,

$$h_1(n) = (-1)^n h_0(N - 1 - n), \quad 0 \le n \le N - 1 \tag{4.83}$$

The coefficients of $h_{11}(n)$ obtained from Table 4.2 and Eq. (4.83) are shown in Table 4.6.

The coefficients of low-pass filters, $g_0(n)$, of imaginary tree after first stage are computed by using half-sample delay condition as given in Eq. (4.49a) and are presented in Table 4.7, and Table 4.8 shows the coefficients of high-pass filter, $g_1(n)$, of imaginary tree which forms the QCF pair with $g_0(n)$.

The above-stated analysis filters are orthogonal because it satisfies the conditions illustrated by above-stated equations, and the corresponding synthesis filter is time-reversed version of analysis filter.

Table 4.7 Coefficients of $g_0(n)$

$g_0(0)$	$g_0(1)$	$g_0(2)$	$g_0(3)$	$g_0(4)$
0.003336	−0.01258	−0.00624	0.077583	−0.03225

$g_0(5)$	$g_0(6)$	$g_0(7)$	$g_0(8)$	$g_0(9)$
−0.24233	0.138449	0.724418	0.60392	0.160127

Table 4.8 Coefficients of $g_1(n)$

$g_1(0)$	$g_1(1)$	$g_1(2)$	$g_1(3)$	$g_1(4)$
0.160127	−0.60392	0.724418	−0.13845	−0.24233

$g_1(5)$	$g_1(6)$	$g_1(7)$	$g_1(8)$	$g_1(9)$
0.03225	0.077583	0.006242	−0.01258	−0.00334

4.7.4 Filters for 2D DT-CWT

Image is a 2D signal, and wavelet decompression of image can be carried out by making use of above-stated filters either by separable DT-CWT or by the development of 2D non-separable DT-CWT that represents edges more efficiently than the former one in $\theta = \{+15°, +45°, +75°, -15°, -45°, -75°\}$ orientations. The non-separable 2D filters are obtained from 1D DT-CWT filters.

Let h_A, h_{LH}, h_{HL}, and h_{HH} are 2D non-separable filters of LL (approximation), LH (horizontal), HL (vertical), and HH (diagonal) sub-bands of real tree obtained from $h_0(n)$ and $h_1(n)$, which are given by Eq. (4.84):

$$\left.\begin{array}{l} h_A = h_{LL} = h_0^T(n) \cdot h_0(n) \\ h_{LH} = h_0^T(n).h_1(n) \\ h_{HL} = h_1^T(n).h_0(n) \\ h_{HH} = h_1^T(n).h_1(n) \end{array}\right\} \qquad (4.84)$$

Similarly, the non-separable filters of size $(N \times N)$ for imaginary tree are given by Eq. (4.75):

$$\left.\begin{array}{l} g_A = g_{LL} = g_0^T(n) \cdot g_0(n) \\ g_{LH} = g_0^T(n) \cdot g_1(n) \\ g_{HL} = g_1^T(n) \cdot g_0(n) \\ g_{HH} = g_1^T(n) \cdot g_1(n) \end{array}\right\} \qquad (4.85)$$

4.7.5 Rotated Complex Wavelet Filters

Kim and Udpa [19] designed new 2D non-separable and rotated discrete wavelet filter (RDWF) for clear characterization of diagonally oriented texture (removal of

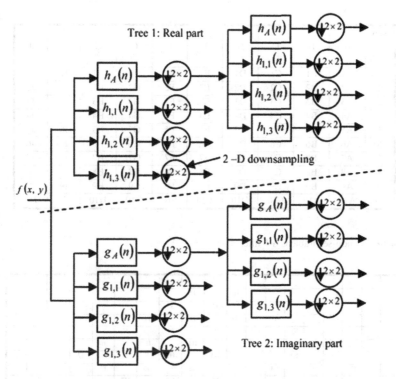

Fig. 4.18 Two levels of decomposition of input image $f(x, y)$ with dual-tree 2D RCWFs [18]

ambiguity of $+45°$ or $-45°$ in diagonal components of DWT) to improve the results of texture image classification.

They proposed rotation of 2D non-separable DWT filter of size $(N \times N)$ by $45°$ to obtain new rotated 2D non-separable filter of size $[(2N - 1) \times (2N - 1)]$. The new rotated filter (RDWF) also retained the orthogonality because it satisfies the condition of orthogonality as illustrated by Eq. (4.86) [19].

$$\frac{1}{2\pi} \int_{\omega=-\infty}^{\infty} S_i(\omega) \cdot \overset{*}{S_j}(\omega)d\omega = 0, \quad i \neq j \tag{4.86}$$

where $S_i(\omega)$ is the Fourier transform of 2D RDWFs and $S_j(\omega)$ is the conjugate of Fourier transform of filter.

The above-stated approach of designing rotated filters is explored by Kokare et al. [18] for designing RCWFs by rotating non-separable 2D DT-CWT filters represented by Eqs. (4.84) and (4.85) for improving the retrieval rate of texture images in the problems of content-based image retrieval. The decomposition of input image with 2D RCWF can be performed by filtering a given image $f(x, y)$ with 2D RCWFs, followed by 2D downsampling operations as illustrated in Fig. 4.18.

The 2D filters are shown in Fig. 4.18, i.e., $h_A, h_{1,1}, h_{1,2}, h_{1,3}, g_A, g_{1,1}, g_{1,2}$ and $g_{1,3}$ are $45°$-rotated versions of the filters that are obtained using Eqs. (4.32) and (4.39). They correspond to $\phi_1, \psi_{1,1}, \psi_{1,2}, \psi_{1,3}\phi_2, \psi_{2,1}, \psi_{2,2}$ and $\psi_{2,3}$, respectively.

A B

Row \ Col	A	B	C	D	E	F	G	H	I	J
1	A1	B1	C1	D1	E1	F1	G1	H1	I1	J1
2	A2	B2	C2	D2	E2	F2	G2	H2	I2	J2
3	A3	B3	C3	D3	E3	F3	G3	H3	I3	J3
4	A4	B4	C4	D4	E4	F4	G4	H4	I4	J4
5	A5	B5	C5	D5	E5	F5	G5	H5	I5	J5
6	A6	B6	C6	D6	E6	F6	G6	H6	I6	J6
7	A7	B7	C7	D7	E7	F7	G7	H7	I7	J7
8	A8	B8	C8	D8	E8	F8	G8	H8	I8	J8
9	A9	B9	C9	D9	E9	F9	G9	H9	I9	J9
10	A10	B10	C10	D10	E10	F10	G10	H10	I10	J10

C D

Fig. 4.19 2D non-separable CWT filter

										J1										
									I1		J2									
								H1		I2		J3								
							G1		H2		I3		J4							
						F1		G2		H3		I4		J5						
					E1		F2		G3		H4		I5		J6					
				D1		E2		F3		G4		H5		I6		J7				
			C1		D2		E3		F4		G5		H6		I7		J8			
		B1		C2		D3		E4		F5		G6		H7		I8		J9		
A•	A1		B2		C3		D4		E5		F6		G7		H8		I9		J10	•C
		A2		B3		C4		D5		E6		F7		G8		H9		I10		
			A3		B4		C5		D6		E7		F8		G9		H10			
				A4		B5		C6		D7		E8		F9		G10				
					A5		B6		C7		D8		E9		F10					
						A6		B7		C8		D9		E10						
							A7		B8		C9		D10							
								A8		B9		C10								
									A9		B10									
										A10										

•
D

Fig. 4.20 2D non-separable RCWF obtained from 2D non-separable CWF

Figure 4.19 shows one of the non-separable 2D DT-CWT filter obtained from any two of four 1D filters. Figure 4.20 illustrates the 45°-rotated filter of it. The size of rotated filter is $[(2N - 1) \times (2n - 1)]$.

The 100 coefficients of non-separable 2D DT-CWT (as illustrated by Fig. 4.20) are placed in appropriate cells of Fig. 4.21 to obtain the respective rotated complex wavelet filter. The black cells of 2D RCWF matrix are filled with zeros. Thus, these $[(2N - 1) \times (2N - 1)]$ size 2D filters are obtained by using each 2D non-separable CWT filters. These filters are oriented in $\theta = \{0°, +30°, +60°, +90°, 120°, -30°\}\}$, which are translated by 45°.

(a)

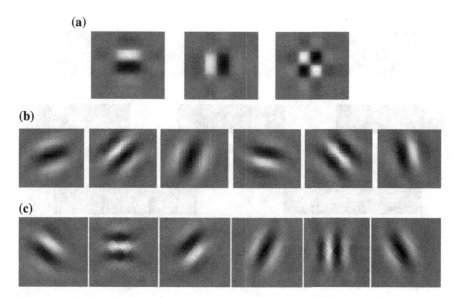

(b)

(c)

Fig. 4.21 Impulse responses of **a** DWT sub-band filters, **b** DT-CWT sub-band filters, and **c** RCWF sub-bands

4.8 Experimental Results

As described earlier, although DWT being orthogonal and implemented efficiently with the help of filter bank approach for MRA, it has oblivious limitations of shift variance and poor directionality. It can capture features only in three orientations, horizontal (0°), vertical (90°), and diagonal (45° or −45°), which is not adequate for analysis and feature extractions of randomly oriented, rich-textured iris images. Therefore, to overcome the limitations of DWT and preserving its advantages over Gabor filters, we designed 2D DT-CWT filter bank, as described in the above section, to capture the iris features in six different orientations along lines at angles of $\theta = \{+15°, +45°, +75°, −15°, −45°, −75°\}$. We also derived 2D non-separable RCWF from 2D non-separable complex wavelet filters, as illustrated in Sect. 4.7 to capture iris features in additional six orientations along lines at angles of $\theta = \{0°, +30°, +60°, +90°, 120°, −30°\}$. These orientations are lagged by 45° from the six orientations of CWT. The impulse responses of newly designed 2D DWT filters, 2D DT-CWT filters, and 2D RCWF are derived and are shown in Fig. 4.21a–c, respectively.

In this study, we decomposed iris image to third level by using standard DWT filter bank, DT-CWT filter bank, and RCWF to capture the iris features in $\theta = \{0°, 90°, ±45°\}, \theta = \{+15°, +45°, +75°, −15°, −45°, −75°\}$, and $\theta = \{−30°, 0°, 30°, 60°, 90°, 120°\}$, respectively. The DT-CWT filter bank and RCWF capture complementary iris features in six orientations each. Therefore, we

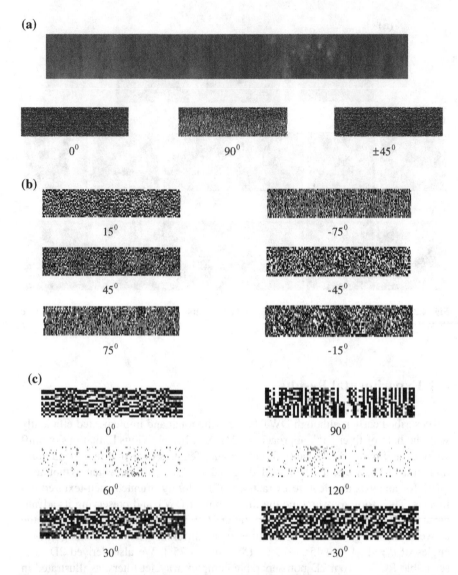

Fig. 4.22 Iris image and its sub-band decomposition using DWT, DT-CWT, and RCWF. **a** Iris image. **b** Sub-bands of DT-CWT. **c** Sub-bands of RCWF

used these filter banks jointly to obtain iris features in 12 different directions. The original iris image and its three sub-bands using DWT, six sub-bands using DT-CWT, and six sub-bands using RCWF are shown in Fig. 4.22.

The performance of iris recognition using DWT, DT-CWT, RCWF, and combination of DT-CWT and RCWF is evaluated on iris images segmented from eye

images of UBIRIS database and UPOL database by using the proposed iris segmentation method as presented in Chap. 3. Each iris image is analyzed up to third
level of decomposition, and four different sets of features are created as given
below:

(a) Feature set 1: DWT only.
(b) Feature set 2: DT-CWT only.
(c) Feature set 3: RCWF only.
(d) Feature set 4: combination of RCWF and DT-CWT.

The feature vectors for each of the above-stated sets are created by computing
energy (E) and standard deviation (σ) of each sub-band at each level. Daugman
[37] reported that energy-based approaches are supported by physiological studies
of the visual cortex. Kokare et al. [38] used combination of these feature parameters for content-based texture image retrieval and reported the improved retrieval
performance. The energy (E_l) and standard deviation (σ_l) of the sub-band are computed as follows using Eq. (4.87):

$$E_l = \frac{1}{M \times N} \sum_{i=1}^{M} \sum_{j=1}^{N} W_l(i, j) \tag{4.87}$$

$$\sigma_l = \left[\frac{1}{M \times N} \sum_{i=1}^{N} \sum_{j=1}^{M} (W_l(i, j) - \mu_l)^2 \right]^{\frac{1}{2}} \tag{4.88}$$

where $W_l(i, j)$ is the lth sub-band of $M \times N$ size and σ_l is the mean of lth sub-
band. A feature vector of using E_l and σ_l and their combination are illustrated by
Eqs. (4.89)–(4.91) feature components.

Using only energy

$$\overline{f_E} = [E_1 E_2 \ldots E_n] \tag{4.89}$$

Using only standard deviation

$$\overline{f_\sigma} = [\sigma_1 \sigma_2 \ldots \sigma_n] \tag{4.90}$$

Using a combination of energy, mean, and standard deviation

$$\overline{f_{E\sigma}} = [E_1 E_2 \ldots E_n \sigma_1 \sigma_2 \ldots \sigma_n] \tag{4.91}$$

The maximum length of feature vector using all types of feature parameter for set
1, set 2, set 3, and set 4 is 18, 36, 36, and 72, respectively, because 9 energies
and 9 standard deviations, 18 energies and 18 standard deviations, and 18 energies and 18 standard deviations are computed for sub-bands of DWT, DT-CWT,
and RCWF at three levels of decompositions. Table 4.9 shows the values of energies and standard deviations of sub-bands of DWT, DT-CWT, and RCWF at single
level of decomposition for the same iris image.

Daugman used Gabor filters to extract phase-based features of iris to generate quantized iris code. Various implementations of his systems are available, and

Table 4.9 Values of energies and standard deviations of sub-bands of DWT, DT-CWT, and RCWF at first level of decomposition of sample iris image

Feature method	E_1	E_2	E_3	E_3	E_5	E_6	σ_1	σ_2	σ_3	σ_4	σ_5	σ_6
DWT	2.85	2.58	1.63	–	–	–	1.10	0.83	0.32	–	–	–
DT-CWT	12.75	0.90	5.88	12.91	0.90	5.80	2.141	0.637	1.477	2.137	0.6368	1.479
RCWF	4.223	4.23	3.09	3.166	1.90	1.361	3.893	6.224	6.173	5.034	7.869	7.239

we used one such implementation to compare the performance of iris recognition using DT-CWT and RCWF (our proposed method) with that using Gabor filters [1–4, 39] (Daugman's method).

The experimental results of this chapter justified the superiority of Canberra distance over Euclidian and Hamming distances for matching the feature vector of query image with feature vectors of stored images. Therefore, in this study, Canberra distance as given by Eq. (4.29a) is used for matching the query image with stored templates.

4.8.1 Role of Energy and Standard Deviation of Sub-bands in Iris Recognition

Energy and standard deviation of wavelet sub-bands have been calculated, and three types of feature vectors are formed as given by Eqs. (4.89)–(4.91). The performance of each type of feature vector is presented in Fig. 4.23, and it is observed that the feature vector created by combination of energies and standard deviations provides much superior recognition performance as compared to feature vectors of individual.

4.8.2 Recognition Performance of Various Feature Extraction Methods

Recognition performance of Iris recognition using Gabor filters (Daugman's method) DWT, DT-CWT, RCWF, and combination of DT-CWT and RCWF are analyzed with the help of ROC and EER.

In Fig. 4.24, ROC curve shown by pink color is obtained by using DT-CWT and RCWF jointly and green-colored curve is obtained by using Gabor filters. These characteristics showed the excellent and comparable recognition performance by both the methods. Our method performed slightly better than Gabor filter method.

FAR and FRR are found out at various normalized values of threshold, and the variation in FAR and FRR against the normalized threshold is plotted to obtain an EER for proposed method as shown in Fig. 4.25.

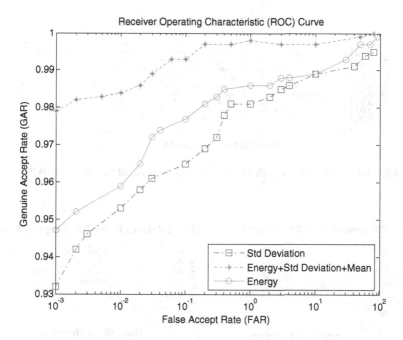

Fig. 4.23 ROC for standard deviation only, energy only, and combination of both

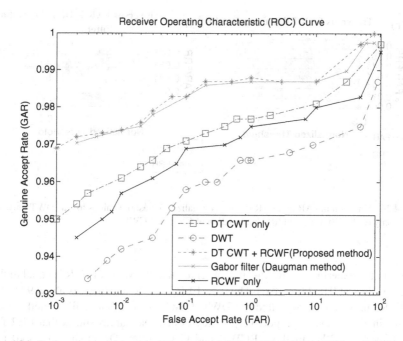

Fig. 4.24 Performance of iris recognition using DWT, CWT, RCWF, CWT + RCWF, and Gabor filters

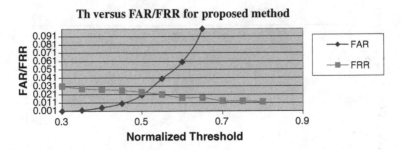

Fig. 4.25 Variation in FAR and FRR for various values of threshold for CWT + RCWF

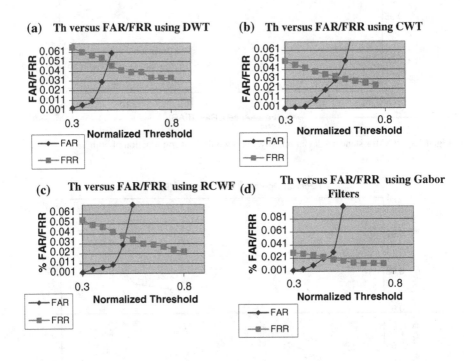

Fig. 4.26 Variation in FAR and FRR versus Normalized Threshold values for (**a**) DWT only (**b**) CWT only (**c**) RCWF only and (**d**) Combination of CWT & RCWF

From the experimental results, it is observed that the value of FAR is equal to the value of FRR at 0.022 (approx.). Thus, the EER is 2.2 % and GAR is about 98 %.

EER curves are also plotted for DWT, DT-CWT, and Gabor filter methods as shown in Fig. 4.26a–d, respectively. Analysis of these curves shows that EER for DWT method, CWT method, RCWT, and Gabor filter (Daugman's method) are 4.8, 3.2, 3.9, and 2.1 %, respectively.

Table 4.10 Comparison of various methods

Parameter method	EER (%)	Average recognition accuracy (%)	Size of feature vector (Bytes)
DWT method	4.8	91.5	18
CWT method	3.2	95.5	36
RCWT method	3.8	93.0	36
Proposed (CWT + RCWT) method	2.2	97.5	72
Gabor filter (Daugman) method	2.1	97.0	256

The performance of iris recognition by using each method in terms of recognition accuracy and EER is summarized in Table 4.10.

These characteristics showed that proposed method and Daugman method outperformed other methods. Moreover, our method performed slightly better than Gabor filter method because the proposed method captures iris features in 12 orientations which are more than its counterpart.

4.8.3 Performance Analysis Using Intra-class and Inter-class Separation Test

ROC curves presented in Fig. 4.24 reveal almost equal results by our method and Daugman method. Therefore, we conducted an experiment to analyze the correct match and nearest non-match distance for correctly recognized test images to compare the performance of our method and Daugman method. The match-normalized distance (intra-class distance) and nearest non-match-normalized distance (closest inter-class distance) for all correctly recognized test images are plotted. Figures 4.27 and 4.28 show the result of this experiment for our method and Daugman method, respectively.

Figures 4. 27 and 4.28 illustrate the wide separation between inter-class and intra-class in case of our system as compared to Daugman method. Thus, our method is less prone to recognition errors as compared to the counterpart in case of variation in texture features due to noise interference or shift variance.

4.8.4 Analysis of Size of Feature Vector and Processing Time

Processing time analysis of all the methods is illustrated in Table 4.11. For this analysis, all the methods are implemented in MATLAB 7.0 and executed on Intel Pentium Mobile, 2.8 GHz, 512 MB RAM, when all non-essential processes were terminated.

Table 4.11 reveals saving of 0.39 s (42 %) in feature extraction using our method when compared to Daugman's method. It also saves 0.16 s (37 %) time in

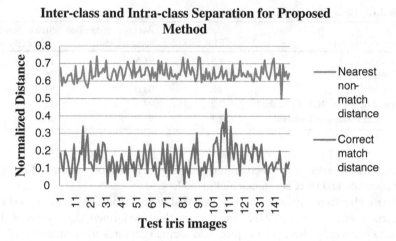

Fig. 4.27 Inter-class and intra-class separation in terms of nearest non-match distance and correct matching distance for proposed method

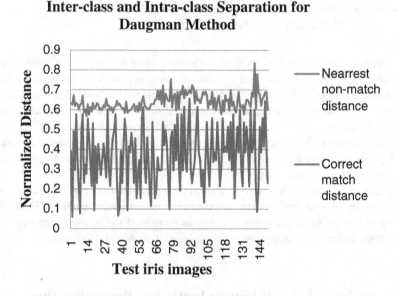

Fig. 4.28 Inter-class and intra-class separation in terms of nearest non-match distance and correct matching distance for Daugman's method

matching because of its small feature vector. Thus, about 80 % time efficiency is achieved by our method when compared to Daugman's method.

However, it may be noted that the values of timings stated in Table 4.11 are just indicative because processing time can only be derived from actual number of

Table 4.11 Comparison of various methods with respect to processing time and length of feature vector

Method performance	DWT	CWT	RCWF	CWT + RCWF (proposed method)	Gabor filter (Daugman method)
Feature vector size(bytes)	18	36	36	72	256
Average processing time for feature vector creation (seconds)	0.3348	0.5704	0.6786	0.9254	1.3206
Average processing for matching test image with 200 classes(seconds)	0.2052	0.3196	0.3114	0.4446	0.6094
Average processing time for our first segmentation method (seconds)	0.4495	0.4495	0.4495	0.4495	0.4495
Average processing time for our second segmentation method (seconds)	1.17	1.17	1.17	1.17	1.17
Average processing time for Daugman segmentation method (seconds)	1.83	1.83	1.83	1.83	1.83
Overall average processing time 1 (2 + 3 + 4)	0.9895	1.3395	1.4395	1.8195	2.3795
Overall average processing time 2 (2 + 3 + 5)	1.71	2.06	2.16	2.54	3.1
Overall average processing time 3 (2 + 3 + 6)	2.37	2.72	2.82	3.2	3.76

Table 4.12 Energies of three-directional sub-bands of DWT (at level 1) for original iris image and shifted iris images

	E^V	E^H	E^D	E^T
Original iris	2.855999	2.58632	1.638185	7.080504
Iris shifted by 5 columns	2.25911	1.562517	1.02923	4.850857
Iris shifted by 10 columns	1.426671	0.966498	0.695289	3.088458
Iris shifted by 15 columns	1.215206	0.705403	0.104255	2.024864

computations involved in the implementation of that algorithm and not by measuring the execution time. But from the basic theories of wavelets and Gabor filter implementations, efficient DWT implementation using filter banks on the one side and the challenges and difficulties in parameterization of Gabor filters on the other side makes wavelet-based approach computationally efficient, which has been reflected in time required to create a feature vector using DWT, CWT, RCWF, CWF + RCWF, and Gabor filter method. Larger size feature vector requires more time for matching. Thus, keeping smaller size feature vector in proposed method reduces the effective processing time. Moreover, processing time can be saved in iris segmentation method as indicated in Table 4.11. Daugman segmentation method requires more time as compared to both of our segmentation methods. As presented in Chap. 3, our first method is most time-efficient, but the recognition rate with this method is less when compared to our second method of iris segmentation. In our study, we used first method for iris segmentation of UBIRIS database and second method for segmentation of UPOL database and combined the iris images of both the databases for evaluating the recognition performance.

4.8.5 Shift Invariance Test of DT-CWT

As stated in the earlier section, shift variance is one of the limitations of DWT, whereas DT-CWT is approximately shift invariant. To justify it, we computed the energies of directional sub-bands of DWT and DT-CWT for original iris image and for iris image with shift of 5, 10, and 15 samples (columns for image).

The energies of 3 sub-bands of DWT and 6 sub-bands of DT-CWT (at level 1 decomposition) of original images and shifted images are listed in Tables 4.12 and 4.13, respectively. An example of shifting of iris image by 10 samples is shown in Fig. 4.29.

Comparative study of above-stated tables reveals over 71.40 and 1.10 % variation in energy of original iris image and 15 column-shifted iris image using DWT and DT-CWT, respectively. This result indicates that DT-CWT is almost shift invariant. Therefore, iris recognition using DWT leads to incorrect recognition if query image is shifted version of stored template of same class, whereas using CWT, the recognition performance is not affected by the shifted query image.

Table 4.13 Energies of three-directional sub-bands of DT-CWT (at level 1) for original iris image and shifted iris images

	E^{75}	E^{45}	E^{15}	E^{-75}	E^{-45}	E^{-15}	E^{T}
Original iris	12.75904	0.906684	5.880034	12.91811	0.902546	5.801106	39.16752
Iris shifted by 5 columns	12.72545	0.905554	5.850361	12.82075	0.889819	5.760591	38.95252
Iris shifted by 10 columns	12.5929	0.897003	5.837254	12.75238	0.892794	5.750614	38.72295
Iris shifted by 15 columns	12.5603	0.897912	5.87357	12.71454	0.894058	5.792703	38.73308

Fig. 4.29 Shifting of iris image by 10 columns. **a** Original iris image. **b** First 10 columns of the iris (iris image portion before shift value, 10 here). **c** Remaining columns of the iris (iris portion after shift value, 10 here). **d** Shifted iris image by 10 columns

4.9 Summary

The following conclusions are drawn from the analysis of above-stated experimental results.

1. Our proposed system using combination of DT-CWT and RCWF decomposed iris image into 12 sub-bands oriented in 12 directions $\theta = (-75°, -60°, -45°, -15°, 0°, 15°, 45°, 60°, 75°)$ as compared to only 3 in case of DWT and 6 each independently by DT-CWT and RCWF. Therefore, randomly oriented texture of iris is very well captured by our method as compared to other methods.

2. ROC curves presented in Fig. 4.24 reveal that iris recognition using combination of DT-CWT is RCWF produced much better performance as compound to that using DWT, DT-CWT, and RCWF. Moreover, performance of our method is marginally better than that of Daugman's method, which is considered as one of the best iris recognition methods among all existing methods proposed by various researchers.

3. Figures 4.25 and 4.26 show that for our method, separation between correct matching distance and nearest unmatched distance is more as compared to that for Daugman's method. This provides robustness to our method because its recognition performance remains unaffected (adversely) even for poor-quality images (noisy images/shifted image) due to problems in imaging system, less cooperative, or untrained users.

4. Table 4.13 shows negligible variations in the values (features) of energies of two iris images which are shifted version of each other, i.e., they are not same in special domain. This is possible because DT-CWT and Gabor filter are shift variant. Daugman has compensated adverse effect of shifts in template-matching stage but at the cost of additional computational overheads.

5. Timing analysis presented in Table 4.11 reveals that our method saved over 70 % of processing time as compared to Daugman's method. Although values indicated in the table are indicative and actual processing time depends on various factors, the computational superiority of our algorithm is justified with the help of following points.

   (a) Our segmentation algorithm requires less computations as compared to Daugman's segmentation algorithm (rubber sheet model) as explained in Chap. 3 of this book.

   (b) DT-CWT is computationally more efficient than Gabor filter because DT-CWT is efficiently implemented using filter bank approach. Moreover, DT-CWT filters are orthogonal or biorthogonal.

6. Figure 4.23 illustrates that energies of wavelet sub-bands can represent iris more correctly than standard deviation, but combination of energies and standard deviation provide much superior results than individual one.

References

1. J. Daugman, High confidence visual recognition of persons by a test of statistical independence. IEEE Trans. Pattern Anal. Mach. Intell. **15**(11), 1148–1161 (1991)
2. J. Daugman, Biometric personal identification system based on iris analysis. U.S. Patent No. 5,291,560 (1994)
3. J. Daugman, The importance of being random: statistical principles of iris recognition. Pattern Recogn. 279–291 (2003)
4. J. Daugman, How iris recognition works. IEEE Trans. Circuits Syst. Video Technol. **14**(1), 21–30 (2004)
5. R. Wildes, Iris recognition: an emerging biometric technology. Proc. IEEE **85**(9), 1348–1363 (1997)

6. H. Sankaran, A. Gotchev, K. Egiazarian, J. Astola, Complex wavelets versus Gabor wavelets for facial feature extraction: a comparative study, in *Proceedings of the conference on image processing: algorithms and systems IV*, vol. 5672 of SPIE, San Jose, Calif, USA (2005), pp. 407–415

7. T. Weldon, W. Higgins, D. Dunn, Efficient Gabor filter design for texture segmentation. Pattern Recogn. **29**(12), 2005–2015 (1996)

8. L. Nanni, D. Maio, Weighted sub-Gabor for face recognition. Pattern Recogn. Lett. **28**(4), 487–492 (2007)

9. W. Choi, S. Tse, K. Wong, K. Lam, Simplified Gabor wavelets for human face recognition. Pattern Recogn. **41**(3), 1186–1199 (2008)

10. D. Liu, K. Lam, L. Shen, Optimal sampling of Gabor features for face recognition. Pattern Recogn. Lett. **25**(2), 267–276 (2004)

11. E. Simoncelli, W. Freeman, E. Adelson, D. Heeger, Shiftable multiscale transforms. IEEE Trans. Inf. Theory **38**(2), 587–607 (1992)

12. J. Lina, Image processing with complex Daubechies wavelets. J. Math. Imaging Vis. **7**, 211–223 (1997)

13. T. Bulow, G. Sommer, Hypercomplex signals: a novel extension of the analytic signal to the multidimensional case. IEEE Trans. Signal Process. **49**, 11 (2001)

14. W. Lawton, Applications of complex valued wavelet transforms to subband decomposition. IEEE Trans. Signal Process. **41**(12), 3566–3568 (1993)

15. I. Selesnick, R. Baraniuk, N. Kingsbury, The dual tree complex wavelet transform: a coherent framework for multiscale signal and image processing. IEEE Signal Process. Mag. 123–151 (2005)

16. I. Selesnick, The design of approximate Hilbert transform pairs of wavelet bases. IEEE Trans. Signal Process. **50**(5), 1144–1152 (2002)

17. N. Kingsbury, The dual-tree complex wavelet transform: a new technique for shift invariance and directional filters, in *Proceedings of the 8th IEEE DSP workshop*, Utah (1998), pp. 86.20

18. M. Kokare, P. Biswas, B. Chatterji, Rotation invariant texture features using rotated complex wavelet for content based image retrieval, in *Proceedings of the IEEE international conference on image processing*, vol. 1, Singapore (2003), pp. 393–396

19. N. Kim, S. Udpa, Texture classification using rotated wavelet filters. IEEE Trans. Syst. Man Cybern. Part: A, Syst. Humans **30**(6), 847–852 (2000)

20. S. Mallat, A theory for multiresolution signal decomposition: the wavelet representation. IEEE Trans. Pattern Anal. Mach. Intell. **11**(7), 674–693 (1989)

21. I. Daubechies, Ten lectures on wavelets. SIAM, CBMS series, Philadelphia (1992)

22. A. Abdelnour, I. Selesnick, Nearly symmetric orthogonal wavelet bases, in *Proceedings of the IEEE international conference on acoustics, speech, signal processing* (2001)

23. R. Wildes, J. Asmuth, S. Hsu, R. Kolczynski, J. Matey, S. Mcbride, Automated noninvasive iris recognition system and method. United States Patent No. 5572596 (1996)

24. R. Wildes, J. Asmuth, G. Green, S. Hsu, R. Kolczynski, J. Matey, S. McBride, A machine-vision system for iris recognition. Mach. Vis. Appl. **9**, 1–8 (1996)

25. Y. Huang, S. Luo, E. Chen, An efficient iris recognition system, in *Proceeding of the international conference on machine learning and cybernetics*, vol. 1 (2002), pp. 450–454

26. Z. Sun, Y. Wang, T. Tan, J. Cui, Cascading statistical and structural classifiers for iris recognition, in *Proceedings of the international conference on image processing* (2004), pp. 1261–1262

27. C. Park, J. Lee, Extracting and combining multimodal directional iris features, in *Springer LNCS 3832: international conference on biometrics* (2006), pp. 389–396

28. I. Daubechies, Orthonormal bases of compactly supported wavelets. J. Commun. Pure Appl. Math. **41**, 909–996 (1988)

29. L. Chenhong, L. Zhaoyang, Efficient iris recognition by computing discriminable textons, in *Proceedings of the international conference on neural networks and brain*, vol. 2 (2005), pp. 1164–1167

30. C. Chou, S. Shih, W. Chen, V. Cheng, Iris recognition with multi-scale edge-type matching, in *Proceedings of the international conference on pattern recognition* (2006), pp. 545–548
31. W. Boles, B. Boashash, A human identification technique using images of the iris and wavelet transform. IEEE Trans. Signal Process. **46**(4), 1185–1188 (1998)
32. G. Strang, T. Nguyen, *Wavelets and filter banks*. Wellesley-Cambridge Press (1996)
33. J. Lina, Image processing with complex Daubechies wavelets. J. Math. Imaging Vis. **7**, 211–223 (1997)
34. M. Vetterli, C. Herley, Wavelets and filter banks: theory and design. IEEE Trans. Acoust. Speech Signal Process. 2207–2232 (1992)
35. I. Selesnick, Hilbert transform pairs of wavelet bases. IEEE Signal Process. Lett. **8**(6), 170–173 (2001)
36. M. Smith, T. Barnwell, Exact reconstruction techniques for tree-structured subband coders. IEEE Trans. Acoust. Speech Signal Process. **34**, 434–441 (1986)
37. J. Daugman, Two-dimensional spectral analysis of cortical receptive field profile. Vision. Res. **20**, 847–856 (1980)
38. M. Kokare, B. Chatterji, P. Biswas, Wavelet transform based texture features for content based image retrieval, in *Proceedings of the 9th national conference on communications*, Chennai, India, pp. 443–447 (2003)
39. L. Masek, Recognition of human iris patterns for biometric identification. Master's thesis, University of Western Australia (2003). Available on: http://www.csse.uwa.edu.au/~pk/stude ntprojects/libor/LiborMasekThesis.pdf

Chapter 5
Conclusion and Future Scope

Abstract Based on the results of experimentations and its analysis, concrete conclusions have been derived in this chapter. Considering the strengths and weaknesses of proposed algorithms of iris segmentation and feature extraction, important future directions also have been presented in this chapter which can be helpful for budding researchers in this filed.

Keywords Iris recognition system • Iris segmentation of realistic images • DT-CWT and RCWF • Accurate iris segmentation • Pupil dynamics • Iris image acquisition • Iris imaging inconsistencies

In today's networked world, the need to maintain the security of information or physical property is becoming both increasingly important and difficult. Most of the times, criminals have been taking the advantage of a fundamental flaw in the conventional access control systems. The access control systems based on biometrics have a potential to overcome most of the deficiencies of current security systems and have been gaining importance in recent years. Comparison of various biometric traits shows that iris is very attractive biometric because of its uniqueness, stability, and non-intrusiveness. It provides automated methods of verifying or recognizing the identity of a living person on the basis of iris texture characteristics.

Number of problems are required to be tackled in order to develop a successful iris recognition system, namely aliveness detection, iris segmentation, and feature extraction. It turns out that an iris is a part of human eye which is surrounded by other parts, namely sclera, pupil, eyelids, and eyelashes. Therefore, accurate segmentation of iris without loss of iris data from a realistic eye is a very important step of any automated iris recognition system. Iris is rich in texture image having randomly oriented multi-directional edges. Therefore, design and implementation of effective feature extraction method decides the success rate of iris recognition system. The system is incomplete if it is not robust against the counterfeit attacks of fake irises.

Issues and challenges related to these three stages of iris recognition system have been addressed in this book.

Two techniques of iris segmentation have been presented in Chap. 3. The first one is fast method of iris segmentation customized for realistic UBIRIS database. This technique has demonstrated fairly good segmentation success rate with very less processing time at the cost of some loss of iris data. This has resulted in fall of recognition performance. Therefore, to improve the recognition performance, the pupil dynamics based novel method for accurate iris segmentation has been developed. The segmentation accuracy of this method is almost 100 % but at the cost of increased processing time as compared to first method. But due to very accurate iris segmentation without any loss, this method has outperformed in terms of recognition accuracy. This method is also capable of accurate iris segmentation of challenging images having low intensity gradients at iris boundaries. For such images, existing methods have underperformed. As this method is based on pupil dynamics, it is inherently used for fake iris detection also.

Thus, for better recognition performance, not only better segmentation rate but accuracy in iris segmentation is also important and segmentation accuracy cannot be compromised at any cost.

The recognition accuracy of iris recognition system is not solely decided by iris segmentation. In fact, it is mainly decided by the success of feature extraction algorithm. This has motivated us to design DT-CWT and RCWF for feature extraction. Feature extraction using combination of DT-CWT and RCWF has decomposed the iris image into 12 sub-bands oriented in 12 directions as compared to only 03 in case of DWT. Therefore, randomly oriented texture of iris is very well captured by this method as compared to other methods. Therefore, this method has performed exceptionally well not only in terms of recognition accuracy but also in terms of computational efficiency.

Following are the major contribution of this research work:

1. This work has addressed the use of DT-CWT and RCWF for iris recognition by designing new filters using Daubechies method.
   (a) The strength of the proposed method of iris recognition using combination of DT-CWT and RCWF lies in its excellent recognition performance (comparable to the Daugman's method) at very less processing time.
   (b) The overall recognition performance of our system is much better than its counterparts, as it showed wider Intra-class and Inter-class separation.
   (c) It has been proved through experiment that DT-CWT is shift invariant and DWT is shift variant.

2. This work proposed the novel iris segmentation method using pupil dynamics.
   (a) Pupil dynamics has been used not only for fake iris detection but also for accurate iris segmentation from realistic eye images without loss/minimal loss of iris data.
   (b) This method is capable of accurate iris segmentation of very challenging images which has very low intensity gradient at the boundaries of iris.
   (c) In this work, unlike others, we have not pre-assumed any shape (circle or ellipse) but obtained the iris of actual shape very accurately.

3. This work proposed the time-efficient iris segmentation algorithm customized for UBIRIS database.
4. Due to very impressive results of iris recognition using DT-CWT and RCWF, we also used DT-CWT for ear recognition to validate the strength of DT-CWT and obtain very satisfying performance.

Following points can be considered for future work:

1. In this research work, image acquisition issues have not been considered, but for any practical recognition system, imaging inconsistencies affect the recognition performance. Therefore, for on field usable system, research on iris image acquisition is recommended for future work.
2. One of the strengths of Daugman's commercial system lies in its compact binary iris code. Therefore, performance evaluation of our system with such quantized binary-coded feature vector can be a work of interest.
3. The robust experimentation with extracting phase information of DT-CWT and RCWF will help to further improve accuracy of iris recognition.
4. The iris recognition with proposed methods for real-time application may be developed by incorporating the suitable image acquisition methodology.
5. Due to very impressive results of iris recognition using DT-CWT and RCWF, its use can be extended for other biometric recognition systems such as face and ear.

About This Book

The book presents three most significant areas in Biometrics and Pattern Recognition. A step-by-step approach for design and implementation of dual-tree complex wavelet transform (DTCWT) and rotated complex wavelet filters (RCWF). In addition to the above, the book provides detailed analysis of iris images and two methods of fast and accurate iris segmentation. It also discusses simplified study of some subspace-based methods and distance measures for iris recognition backed by empirical studies and statistical success verifications.

R. M. Bodade and S. N. Talbar, *Iris Analysis for Biometric Recognition Systems*,
SpringerBriefs in Computational Intelligence, DOI: 10.1007/978-81-322-1853-1,
© The Author(s) 2014

Printed in the United States
By Bookmasters